Taking Private Use Seriously

European University Studies

Europäische Hochschulschriften

Publications Universitaires Européennes

Series II	**Law**
Reihe II	Rechtswissenschaft
Série II	Droit

Volume/Band **5499**

Zhong Zheng

Taking Private Use Seriously

A Critical Evaluation of the Legal Treatment of Private Use under Chinese Digital Copyright Law

PL ACADEMIC RESEARCH

Bibliographic Information published by the Deutsche Nationalbibliothek
The Deutsche Nationalbibliothek lists this publication in the Deutsche Nationalbibliografie; detailed bibliographic data is available in the internet at http://dnb.d-nb.de.

ISSN 0531-7312
ISBN 978-3-631-64188-0 (Print)
E-ISBN 978-3-653-03429-5 (E-Book)
DOI 10.3726/978-3-653-03429-51

Internationaler Verlag der Wissenschaften
Frankfurt am Main 2013

PL Academic Research is an Imprint of Peter Lang GmbH.
Peter Lang – Frankfurt am Main · Bern · Bruxelles · New York · Oxford · Warszawa · Wien

www.peterlang.de

For Fei
and my parents

Acknowledgments

It gives me great pleasure to thank all those whose generous support made the writing and publication of this book possible. The book draws on my Ph.D. dissertation which I defended at Kyushu University. Here, I would like to thank my supervisors professor Ryu KOJIMA and professor HAN Sang Hee, for their confidence in me and for their guidance in leading me to tackle this difficult topic, and especially their help in overcoming a bottleneck midway through the research. Professor KOJIMA enlightened me on a lot of things during "brainstorm" sessions and professor HAN constantly encouraged me to overcome my frustration at times. Without their support and trust, I could hardly have finished the research.

Many thanks to professor Mark FENWICK, for his acute insights regarding the manuscript structure, and his great patience in reviewing my conference paper. I am grateful to professor Shinto TERAMOTO for teaching me about the social network methodology and the dichotomy between fact and assumption. I am also grateful to professor Steven Van UYTSEL who always offered me counterarguments urging me to include a comprehensive analysis, and corrected me about the citation format.

In addition, I wish to thank all faculty members who provided me stimulating questions and helpful insights during the Comprehensive Research Seminars, especially professors Toshiyuki KONO, Shinichi AGO, Michiaki ABE, Caslav PEJOVIC, Mariko IGIMI-YASUTAKE, Machiko HACHIYA, and also Antonio FORMACION. Thanks to professors Peter DRAHOS, Peter K. YU, and Thomas HOEREN, for their valuable comments on my presentation at the 2010 conference "New Spaces, New Actors and the Institutional Turn in Contemporary Intellectual Property Law" jointly organized by Kyushu University and Hokkaido University. Thanks to professor Yoshiyuki TAMURA who invited me to present at the 2010 conference "Multi-Agential Governance and the Idea of Intellectual Property Between Market and Other Social Values" organized by Hokkaido University. Thanks to professors Branislav HAZUCHA, Barton BEEBE, Graham DUTFIELD, Nari LEE, Nobuko KAWASHIMA, Tatsuo TANAKA, Masabumi SUZUKI, Iwao KIDOKORO, Takakuni YAMANE and Susumu MORIMURA for their helpful comments.

Furthermore, I am especially grateful to Ai NAGAO and all the administrative division of the Kyushu University Faculty of Law, as well as Shiuko NISHIDA, Yumi KIMURA, and Chikako YOSHINO from the JICA Kyushu Office. I also owe much to my LL.D. colleagues who gave me incisive suggestions for improvements, especially Sean MCGINTY who proofread the whole manuscript and saved me from many mistakes. Finally, I owe special thanks to my editor Ute WINKELKOETTER at Peter Lang GmbH for helping me prepare the book for publication.

My greatest debt is to my family. My parents have unconditionally supported me in the pursuit of my dreams. When my father was sick, my mother took care of him by herself in order to let me concentrate on my studies. My husband, Fei, tolerated my absence in Japan for three years shortly after we were married so that I could conduct this research, and always supported me with the deepest of love.

As professor KOJIMA has always reminded me, a doctoral research is a small window from which we can see the whole world. The knowledge I have learned and the life I have experienced in the past three years in Kyushu University are an invaluable treasure that I will cherish forever, and their completion marks the beginning of a new journey.

Table of Contents

Abbreviations

AUSFTA	Australia-United States Free Trade Agreement
CAVCA	China Audio-Video Collective Administration
CDPA 1988	UK Copyright, Designs and Patents Act 1988
CNNIC	China Internet Network Information Center
CPC	Communist Party of China
CPCC	Copyright Protection Center of China
DMCA	Digital Millennium Copyright Act of 1988 (US)
DRM	Digital Rights Management
ECC	United Nations Convention on the Use of Electronic Communications in International Contracts (Electronic Communications Convention)
ECHR	European Convention on Human Rights
EU	European Union
IFPI	International Federation of the Phonographic Industry
InfoSoc Directive	Directive 2001/29/EC of the European Parliament and of the Council of 22 May 2001 on the Harmonisation of Certain Aspects of Copyright and Related Rights in the Information Society
IP	Intellectual Property
IPRs	Intellectual Property Rights
ISP	Internet Service Provider
MCSC	Music Copyright Society of China
MPAA	Motion Picture Association of America
NCAC	National Copyright Administration (PRC)
NPC	National People's Congress (PRC)
PPA	General Administration of Press and Publication of the PRC

PRC	People's Republic of China
RIAA	Recording Industry Association of America
RMB	Renminbi (Chinese yuan)
R&D	research and development
SAIC	State Administration for Industry & Commerce (PRC)
SARFT	State Administration of Radio Film and Television (PRC)
SIPO	State Intellectual Property Office (PRC)
TPMs	Technological Protection Measures
TRIPS	Agreement on Trade-Related Aspects of Intellectual Property Rights
WCT	WIPO Copyright Treaty
WIPO	World Intellectual Property Organization
WPPT	WIPO Performances and Phonograms Treaty
WTO	World Trade Organization

Introduction

Private use is full of uncertainty both in theory and in application. It would not be an exaggeration to say that private use is one of the most difficult issues in the copyright field to tackle. It looks trivial and marginal and people usually take it for granted. Given the fact that the copying itself, whether in a private use context or a piracy context, is physically the same, it is very easy to confuse private copying with piracy. It is also a fact that a particular instance of private copying might itself substitute for the purchase of a copy and thus diminish the income of the copyright holder. On the other hand, however, the two are different in essence. Private copies are used for non-commercial private purposes and limited to the private sphere, while pirated copies enter into public distribution and are normally for commercial exploitation. The difficulty, however, is in how to separate them in practice.

In addition, private use is a very important issue which refers to the backbone of the copyright system. How to keep a delicate balance between the interests of copyright holders in the exploitation of their copyrighted works and the benefits of users in the enjoyment of these works is the everlasting fundamental task of copyright law. The private use issue actually deals with the individual users' legitimate interests in using copyrighted works. Regrettably, at present there is not a flourishing field of literature focusing on this specific subject like there is in other areas of copyright, especially in the Chinese context, partly because of the difficulty surrounding it. The challenge of exploring the issues remains enticing to me despite the difficulties and uncertainty that go along with it.

This book explores the place of private use in a digital copyright regime through a critical evaluation of the legal treatment of private use under Chinese digital copyright law.[1] More specifically, it examines the problems of the current legal treatment, as well as their causes and possible solutions.

1 Admittedly, there is no positive law explicitly entitled "the Chinese digital copyright law". I use this expression to denote the aggregation of legislative regulations concerning copyright in the digital environment.

I. Research Content: Private Use

The research content of this book is private use. This section begins with a description of the choice of terminology, the definition of private use, a review of its boundaries and a discussion of the intersection between private use and copyright.

1. Terminology

As Justice Stevens commented in the Sony case,[2] it would plainly be unconstitutional to prohibit a person from singing a copyrighted song in the shower or jotting down a copyrighted poem he hears on the radio.[3] Reading books, listening to music, watching movies or television programs, playing online games…. These are indispensable elements of our study, entertainment and daily life, and the origin of acquirement of knowledge, progress and happiness.

There are many terms which denote these kinds of individual behaviors, such as personal use,[4] public use,[5] individual use, private use, noncommercial use[6] or use by a consumer. In this book, I am consciously choosing the term "private use" rather than a more neutral term such as "individual use", "noncommercial use" or "use by a consumer" in part because I want to highlight two notions – use by individuals and use within the private sphere. In this sense, the term "private use" seems to more appropriately capture these meanings than any other terms, including the commonly used term "personal use".

2 *Sony Corp. of Am. v. Universal City Studios, Inc.*, 464 U.S. 417, 420 (1984).

3 See Jessica Litman, *The Sony Paradox*, 55 CASE W. RES. L. REV. 917, 930–34 (2005) (discussing Justice Stevens' first draft of his opinion in the Sony case).

4 See e.g., Jessica Litman, *Lawful Personal Use*, 85 TEX. L. REV. 1871(2007); Deborah Tussey, *From Fan Sites to Filesharing: Personal Use in Cyberspace*, 35 GA. L. REV. 1129 (2001).

5 See, e.g., Paula Baron, *The Moebius Strip: Private Right and Public Use in Copyright Law,* 70 ALB. L. REV. 1227, 1229 (2007) (The term "public use" refers to all uses by non-owners. Thus, it includes not only transformative uses and public good uses, such as uses for educational purposes, but also some private consumptive uses.).

6 See Neil Weinstock Netanel, *Impose a Noncommercial Use Levy to Allow Free Peer-to-peer File Sharing*, 17 HARV. J. L. & TECH. 1 (2003).

2. Definition

Strictly speaking, there is no solid definition of private use. Copyright laws in most countries only regulate private use explicitly[7] or implicitly[8] within copyright limitations, or through even more abstract policies or principles of copyright.[9] The connotation of private use remains fuzzy and contested as well. In the broad sense, it means the consumption or adaptation of intellectual properties by individual users for their own purposes, including the uncompensated sharing of those works with others.[10] In the narrow sense it means "[p]ersonal private use is that which occurs within you [sic] immediate biological family . . . ."[11]

Standing in a relatively neutral position, "'personal use' is a use that an individual makes for herself, her family, or her close friends". It includes lawful and infringing behavior as well as some controversial ones.[12] However, from the perspective of user's rights, there is no infringement of copyright for the personal use defined as the private use of a work for one's own learning,

7 Continental countries usually contain a general exemption for private copyright, see, e.g., Article L122-5, "Code de la propriété intellectuelle," J.O. 3 July 1992, as last amended by Laws Nos. 94-361 of May 10, 1994, and 95-4 of Jan. 3, 1995 (France), translated in INDUSTRIAL PROPERTY AND COPYRIGHT: MONTHLY REVIEW OF THE WORLD INTELLECTUAL PROPERTY ORGANIZATION (hereinafter IPC WIPO) (September 1995); Article 53 "Urheberrechtsgesetz" BGG1 I pg. 1,273, as last amended by the law of July 19, 1996 (Germany), translated in IPC WIPO (April 1997); Article 31 "Ley de propiedad intelectual," B.O. 17 Nov.1987, as last amended by law No. 43/1884 of Dec. 30, 1994 (Spain), translated in IPC WIPO (October 1995).

8 E.g., 17 U.S.C. § 1001. See Jane C. Ginsburg, *Authors and Users in Copyright*, 45 J. COPYRIGHT SOC'Y U.S.A. 1, 11 (1997) ("The U.S. Copyright Act does not contain a general exemption for private copying. There is a specific exemption/levy statutory scheme for digital audio equipment and media, which includes an exemption for private copying of analog sound recordings. ").

9 See L. RAY PATTERSON & STANLEY W. LINDBERG, THE NATURE OF COPYRIGHT: A LAW OF USERS' RIGHTS 193-94 (Athens: The University of Georgia Press 1991) ("The problem with the law of personal use derives in part from the design of the 1976 act – exclusive rights subject to limitations – which appears to abrogate the law of personal use. Moreover, the law of personal use exists not in an explicit statutory rule but rather by virtue of the politics and principle of copyright.").

10 See Tussey, *supra* note 4, at 1134.

11 Permissions and Copyright of Lutheran Hymnals, available at http://www.lutheran-hymnal.com/about_us/permision_and_copyright.html (last visited Dec. 20, 2010).

12 Litman, *supra* note 4, at 1894.

enjoyment, or sharing with a colleague or friend – without any motive for profit.[13]

Before moving into more controversial territory, I prefer to define private use in this book as the use of works by individual users for their own study, research, or self-entertainment purposes.

In light of the aforementioned definition, only natural persons and not corporations or organizations can become the subject of private use. The purpose of use must, after all, be for personal purposes. When deciding the purpose, should a subjective or an objective approach be adopted? In other words, should the test for determining the purpose of the use be decided according to the subjective intentions of the user or according to objective criteria? As Lionel Bently and Brad Sherman point out:

> At first glance the language of the statute seems to suggest that the test for determining the purpose of the dealing should be decided according to the subjective intentions of the alleged infringer. However, as Aldous LJ pointed out in *Hyde Park*, when deciding the purpose of the dealing, it is not necessary "for the court to put itself in the shoes of the infringer of the copyright". Instead, a more objective approach ought to be adopted. This ensures that the court does "not give any encouragement to the notion that all that is required is for the user to have sincere belief, however misguided, that he or she is criticizing a work or reporting current affairs".[14]

Therefore, when determining the purposes of using copyrighted works by individual users, a more objective standpoint of interpretation should be used.

3. The Boundary between Private Use and Copyright: Enjoyment vs. Exploitation

An individual's use of a copyrighted work for enjoyment is a private use, not subject to exclusive copyright restraints. The distinction between the enjoyment of works and the exploitation of works draws the boundary between the spheres of private use and copyright.

The nature of copyright is a limited monopoly on the marketing of works. In other words, copyright refers to the exploitation of works in the market place. The economic rights granted to copyright holders are therefore also known as

---

13 PATTERSON & LINDBERG, *supra* note 9, at 193.

14 LIONEL BENTLY & BRAD SHERMAN, INTELLECTUAL PROPERTY LAW 193 (2nd ed., Oxford: Oxford University Press 2004).

exploitation rights.[15] Accordingly, the right holder "has the sole right to exploit his work under exploitation rights and to exclude third parties from the exploitation of the work. He should be able to determine whether, when, and how his work will be used to secure the economic exploitation of his work."[16]

In contrast, the use of works by end users makes use of the enjoyment function of works rather than their monetary exploitation merits. According to Kant, enjoyment is a pleasurable sensation.[17] "Enjoyment is pleasure through the senses, and what delights the senses is called agreeable".[18] Through reading a book, watching a movie, listening to music, or other ways of using a work, an individual experiences sensory pleasure. And the enjoyment of works is a natural capacity which human beings possess and of which they can not be deprived. In Aristotle's view, "what we enjoy is the unimpeded exercise of natural capacities, and that enjoyment is the awareness of that exercise".[19]

Generally speaking, there is a distinction between the copyrighted works and the copyright. For example, copyright is only protected within a limited statutory duration. However, the work still exists after the copyright expires and enters into the public domain. Another instance lies in the fact that the transfer of the ownership of a work does not affect its copyright.[20]

Pursuant to the distinction between the works and the copyright, the purchaser of a work only obtains its property rights and not its copyright. The purchaser's property rights in the work are basically confined to private enjoyment.[21] As Benjamin Kaplan emphasizes, it is "fundamental that 'use' is

---

15 See KEY ASPECTS OF GERMAN BUSINESS LAW: A PRACTICAL MANUAL 189-91 (4th ed., Michael Wendler, Bernd Tremml, & Bernard Buecker eds.Verlag: Springer 2008) (stating under the German Copyright Act, "[e]xploitation rights grant exclusive rights to the author for the economic exploitation of the work". These exploitation rights include the right of reproduction, the right of distribution, the right of exhibition, the right of making available a work to the public, the broadcasting right, the right of performance by means of picture or sound carriers, and the right of performance of radio broadcasts.).

16 *Id.* at 190.

17 RACHEL ZUCKERT, KANT ON BEAUTY AND BIOLOGY: AN INTERPRETATION OF THE *CRITIQUE OF JUDGMENT* 240 (UK: Cambridge University Press 2007).

18 Paul Guyer, *Anthropology*, VII: 230, quoting ZUCKERT, *id.* at 239.

19 C.C.W. Taylor, *Pleasure: Aristotle's Response to Plato*, in PLATO AND ARISTOTLE'S ETHICS 13 (Robert Heinaman ed., England: Ashgate Publishing Limited 2003).

20 See e.g., *Stephens v. Cady*, 55 U.S. (14 How.) 528, 531 (1852) (the US Supreme Court stating that "the property in the copy-right is regarded as a different and distinct right, wholly detached from the manuscript or any other physical existence, and will not pass with the manuscript unless included by express words in the transfer.").

21 See GUSTAVO GHIDINI, INTELLECTUAL PROPERTY AND COMPETITION LAW: THE INNOVATION NEXUS 58 (UK: Edward Elgar Publishing Limited 2006).

not the same thing as infringement".[22] "[A]n infringer infringes only the copyright, not the work, and that one can use the work without infringing the copyright."[23] Therefore, the difference between enjoyment and exploitation separates private use from copyright.

This dichotomy has been confirmed by courts for a long time. As Jessica Litman observes, "[t]he case explored … revealed that courts have sought to protect copyright liberties by interpreting copyright law to draw a distinction between exploitation and enjoyment of a copyrighted work. The law, they insisted, gave copyright owners exclusive rights to control the former, but not the latter."[24]

However, it is not easy to draw a clear line between enjoyment and exploitation. "The line … has never been completely obvious, and it has gotten more difficult to discern as networked digital technology enables ordinary people to engage in acts of mass dissemination at negligible expense."[25] Certain sub-factors need to be considered when determining the boundary between enjoyment and exploitation.[26] Nevertheless, "[t]here's broad consensus that unauthorized enjoyment is not and should not be illegal unless it crosses the line into exploitation or otherwise interferes with copyright owners' opportunities to exploit their works."[27]

4. The Intersection between Private Use and Copyright: Copying of Works

The division between the entitlement of users to the enjoyment of works and the right of copyright holders to control the exploitation of their works constitutes the boundary between private use and copyright.[28] Therefore, copyright law conventionally pays little attention to private use. It is only when one wishes to make a copy of a work for a private use purpose that the issue begin to come

22 BENJAMIN KAPLAN, AN UNHURRIED VIEW OF COPYRIGHT 57 (NY: Columbia University Press 1967).

23 PATTERSON & LINDBERG, *supra* note 9, at 199.

24 Litman, *supra* note 4, at 1904.

25 *Id.* at 1908.

26 These sub-factors are commercial use or noncommercial use, and market place or private domain. See *infra* sections 3.1 & 5.1.

27 Litman, *supra* note 4, at 1908.

28 See *id.* at 1904.

into focus.[29] In other words, the copying of works constitutes the intersection between private use and copyright.

In light of the aforementioned distinction between the copyrighted works and the copyright, private copying, namely copying for a private use purpose, is properly classified as the use of a work and serves to enhance the user's convenience in enjoying the works. As Ray Patterson points out:

> One who copies – or makes another use of – a work for a non-market purpose (such as teaching, scholarship, or research) uses the work; one who copies – or makes another use of – a work for the market uses the copyright; The point is that a purchaser pays for the use of the work when he or she buys a copy of it.[30]

In comparison, the reproduction right of a copyright is the first and most well-known right given to copyright holders. It is the right to copy the work.[31] Copyright holders have the exclusive right to make copies of their works. Copying made by others is a violation of the right holder's reproduction right unless the copying is permitted or justified by fair use or other copyright limitations.

It is easy to draw a line between copying for private use and market use in principle. Given that the copying itself is physically or qualitatively the same regardless of whether it is done for private use or for piracy it is, however, very easy to confuse private copying with piracy. Among the various categories of private uses, private copying is thus one of the most controversial issues in copyright debates.

Looking at the issue from the point of view of the historical treatments of private use, the attitudes of copyright law have shifted alongside technological developments. In the analog world, private copying was perceived as non-infringing because it was *de minimas*.[32] Early laws on copyright were not concerned with the kind of small-scale hand-made reproduction that occurred in homes. Even if it is true that private copying is technically infringing, it is

---

29 See PATTERSON & LINDBERG, *supra* note 9, at 193 ("Obviously, there is no infringement of copyright when one reads a copyrighted book, sings a copyrighted song in the shower, watches a copyrighted movie on television, and so on. Only when one wishes to make a copy of such a work for personal use does the issue begin to take focus.").

30 L. Ray Patterson, *Regents Guide to Understanding Copyright and Educational Fair Use*, University of Georgia school of law, Scholarly Works, Paper 349, p.277 (1997), available at http://digitalcommons.law.uga.edu/fac_artchop/349 (last visited Oct. 10, 2010).

31 See BENTLY & SHERMAN, *supra* note 14, at 132.

32 See Pierre N. Leval, *Nimmer Lecture: Fair Use Rescued*, 44 UCLA L. REV. 1449, 1457-58 (1997).

usually too expensive and complicated to effectively prohibit such behavior. This is the "market failure" justification for private copying, the main idea of which lies in the argument that the transaction costs of enforcement will exceed the value of any remedies or licenses.[33] Conventionally, private copying by way of hand copying or typewriting has been the subject of a copyright exemption which is beyond the control of right holders.

With the rise of photocopying and analogue recording equipment in the early 1950s, individuals were able to make relatively easy and cheap copies of works. In order to compensate copyright holder's economic interests affected by private copying, in the 1960s Germany introduced the first copyright levy system in the world.[34] This would become a model for legislators elsewhere.[35]

In general, a private copying levy is a government-mandated scheme in which a special tax or levy is charged on purchases of recordable media. It is a compromise between right holder's monopolistic exploitation interests and user's convenience for enjoyment. Consequently, private copying is legally

---

33 See Wendy Gordon, *Fair Use as Market Failure: A Structural and Economic Analysis of the Betamax Case and Its Predecessors*, 82 COLUM. L. REV. 1600 (1982).

34 The Germany Copyright Act of 1965 included a levy on the sale of sound and video recording equipment. In 1985, the German Copyright Act was modified in two important aspects: (1) concerning sound and audiovisual recordings, a levy was introduced on blank tapes, in addition to the long-standing levy on the sale of recording equipment; and (2) concerning reprographic copies, a levy on photocopying equipment was introduced in connection with a statutory remuneration for every page copied for personal or other private purpose from a work protected by copyright. The latter modification ended the regime that had applied until then, according to which it was permissible to make individual copies of a work for personal and, under certain circumstances, for other private purposes, without payment of remuneration. For a comprehensive discussion, see MARTIN SENFTLEBEN, COPYRIGHT, LIMITATIONS AND THE THREE-STEP TEST: AN ANALYSIS OF THE THREE-STEP TEST IN INTERNATIONAL AND EC COPYRIGHT LAW 53-58 (Hague: Kluwer Law International 2004).

35 Today, most continental European countries have followed the German model and set up the levy system to compensate right holders for private copying, either under the home taping regime or the reprography regime. "The twelve EU Member States that have put in place such a levy system are in chronological order since 1965: Germany (1965), Sweden (1973), Netherlands (1977), Finland (1978), Denmark (1980/educational and 1987/business), France (1988), Spain (1988), Italy (1993), Belgium (1994), Austria (1996), Greece (1996), and Portugal (1998). Ireland, Luxembourg, and the United Kingdom have implemented no levy system to compensate copyright holders for reprographic activities." See Lucie Guibault, *The Reprography Levies Across the European Union*, available at http://www.ivir.nl/publications/guibault/introconcl.htm#ref2 (last visited May 10, 2009).

allowed in many jurisdictions in exchange for the payment of a levy on copying media.[36]

II. Research Context: China

The research context of this book focuses on China. In this section, I will first indicate the universal problems of private use and China's special instance. Then, I will highlight the significance of research on private use in the Chinese context.

1. Universal Problems in Private Use Issues and China's Special Instance

Currently the main problem regarding private use is the conflict between private use and copyright protection by technological protection measures (hereinafter TPMs).[37]

In the new digital environment, the balance between the copyright holder's market exploitation and the user's enjoyment is once again being disrupted. More and more individuals are becoming the target of copyright infringement litigation.[38] When users enjoy the great advantages of digital networks, especially the instant exchange and simultaneous enjoyment of works by the masses through peer-to-peer file sharing, they are potentially becoming online pirates. The explosion in the number of unlicensed transmissions of copyrighted works by end users through file sharing services has come to constitute a formidable threat to the conventional commercial exploitation of their works by

---

36 For example, the US Audio Home Recording Act (AHRA) gives consumers broad rights to copy recorded music for private use (17 U.S.C. §§ 1001-1010). The statute provides for these permissible uses in exchange for a royalty payment for blank digital recording media (17 U.S.C. §§ 1003, 1008).

37 As Dellit and Kendall point out, TPMs can take a variety of forms, including password protection, encryption, or the copy control technology commonly used on CDs or DVDs. See Elouise Dellit & Christopher N. Kendall, *Technological Protection Measures and Fair Dealing: Maintaining the Balance Between Copyright Protection and the Right to Access Information*, 2003 DIGITAL TECH. L.J. 1, para. 5, available at http://beta.austlii.edu.au/au/journals/DTLJ/2003/1.html (last visited Nov. 8, 2010).

38 E.g., the Recording Industry Association of America (RIAA) has already sued over 18,000 individuals for sharing songs online. See Jefferson Graham, R*IAA chief says illegal song-sharing "contained"*, USA TODAY, June 12, 2006, available at http://www.usatoday.com/tech/products/services/2006-06-12-riaa_x.htm (last visited Feb. 2, 2009).

copyright holders.[39] The conventional distinctions between enjoyment and exploitation are becoming blurred.

In response, copyright industries have lobbied legislators to expand their exclusive rights into the digital network and obtain a new "right of making available" to cover the offering and diffusion of works, performances, and phonograms on the Internet.[40] Moreover, copyright holders deploy TPMs to control access and block unauthorized copying of their works while forbidding users to circumvent them.[41] The perfect control by TPMs has sparked fierce debates on the status of private use in the digital environment.

Basically speaking, these arguments can be categorized into two kinds.[42] The first one is associated with the right holders' side and argues that access control by TPMs is important and necessary as the exploitation of works has shifted from having copies to experiencing works.[43] By virtue of TPMs' access control and price discrimination, users only need to pay certain fees depending on the type of use they wish to put the work to. Moreover, private copying is no longer at the inoffensive *de minimas* level, but is now at a scale that constitutes competitive exploitation. Combined with the fact that the market failure justification has also already lost its importance, private copying should therefore no longer be tolerated in digital environment and any unauthorized copying should be considered piracy.

In comparison, another group of commentators argue in favor of the users' benefits. They criticize TPMs control on the use of works exceeding copyright. Additionally they note that the perfect control by TPMs forms a "panoptical prison" which enables copyright holders to monitor, record, and restrict what people read, listen, or watch in the private sphere. This control greatly harms user's fundamental rights, including user's autonomy in the use of works and the

---

39 E.g., RIAA quotes an analysis by the Institute for Policy Innovation which concludes that global music piracy causes $12.5 billion of economic losses every year, 71,060 US jobs lost, a loss of $2.7 billion in workers' earnings, a loss of $422 million in tax revenues, $291 million in personal income tax and $131 million in lost corporate income and production taxes. RIAA, *Piracy: Online and on the Street*, available at http://www.riaa.com/physicalpiracy.php (last visited Jan. 5, 2010).

40 See Article 8 of the WCT, Articles 10, 14 and 15 of the WPPT.

41 The WCT and the WPPT contain obligation concerning technological measures. See Article 11 of the WCT and Article18 of the WPPT.

42 For more details see *infra* sections 3.3 & 3.4.

43 See Jane C. Ginsburg, *From Having Copies to Experiencing Works: The Development of an Access Right in U.S. Copyright Law*, 50 J. COPYRIGHT SOC'Y U.S.A 113, 120 (2003).

right to privacy.[44] Furthermore, they argue that copy-preventing control by TPMs undermines the free speech value of private copying.[45]

In respect of China, conventionally private use has possessed a statutory "free zone" under Chinese copyright law. Private uses, including private copying, belong to the category of fair use exemptions to copyright. Users can use a work for private purposes without authorization from or payment to copyright holders. However, in the Chinese digital copyright regime, there is neither a private use exemption within the fair use of information network transmission right, nor a private use exception for the anticircumvention provisions of TPMs. In this book, I intend to investigate the reasons why such treatment has been adopted and to provide possible solutions to the problems that exist.

2. The Significance of Research on Private Use in the Chinese Context

Arguably, research on the issue of private use in the Chinese context possesses particular importance, not only because the affair itself is full of uncertainty and controversy, but also because it closely relates to the fundamental framework of the copyright regime as well as to the piracy dilemma in China.

First, from the viewpoint of the present legal treatment of private use in the digital environment, China is inclined towards the aforementioned first argument which supports strong copyright protection by TPMs. However, does the operation of this treatment adequately deal with the threats to users' fundamental rights as referred to in the second category of arguments? Which criterion should be followed for decision-making?

Comparative study shows that some other countries have not chosen the path of providing as expansive a protection to TPMs as China has. For example, despite much criticism, the Digital Millennium Copyright Act of 1998 (DMCA) at least includes saving clauses on TPMs protection. Together with judicial restraint on the application of anticircumvention rules, the US copyright system

---

44 See Julie E. Cohen, *A Right to Read Anonymously: A Closer Look at "Copyright Management" in Cyberspace*, 28 CONN. L. REV. 981 (1996) (discussing readers' autonomy right to read anonymously); Julie E. Cohen, *Copyright's Public-Private Distinction*, 55 CASE W. RES. L. REV. 963 (2005) (indicating the significance of user privacy to copyright law).

45 See Rebecca Tushnet, *Copy This Essay: How Fair Use Doctrine Harms Free Speech and How Copying Serves It*, 114 YALE L. J. 535 (2004) (discussing the contribution of private coping to free speech).

seeks a balance between copyright protection and the public interest.[46] Similarly, through judicial interpretation, the Supreme Court of Canada has affirmed "user rights" as a concept integral to copyright law and that user rights are as central to copyright law as author rights.[47] These practices might help us to rethink whether the legal treatment of private use within the TPMs framework in China is proper or not.

Last, one possible criticism might be that even if other countries such as the US and Canada have not adopted very strong protection towards TPMs, this fact alone does not mean China should not do so. Rampant piracy requires strong copyright protection by TPMs in China. This assumption, however, is based on the premise that unauthorized private use is piracy. Is it true that private use equals piracy and that it is therefore necessary to abandon the private use exemption in the digital environment because it is contributing to the spread of piracy in China? If we believe that this is not the case, then what is the relationship between private use and piracy? And what are the real causes of China's piracy dilemma?

III. Methodology and Structure

This book aims at contributing to the debate on the private use issue, particularly from a Chinese perspective. Through a critical evaluation of the legal treatment of private use under Chinese digital copyright law, I intend to explore the place of private use in the digital copyright regime.

Although interesting and relevant to my topic, several questions remain outside the scope of my study. First, transformative use and active use by users which are productive in nature and relate to the exploitation of works are beyond the scope of my research. Generally speaking, there are three categories of using copyrighted works by users: consumptive use, transformative use, and active use.[48] Given the fact that transformative use and active use might also involve

---

46 See *MGE UPS Systems, Inc. v. GE Consumer and Indus., Inc.*, 2010 WL 2820006 (C.A.5 (Tex.)). For more discussion, see *infra* sections 3.2.2 & 3.2.3.

47 See *CCH Canadian Limited v. Law Society of Upper Canada*, 2004 SCC 13. For more discussion, see *infra* section 3.2.4.

48 See Ginsburg, *supra* note 8, at 3 (stating that use by users may be classified into consumptive use and transformative use based on the nature of use); Joseph P. Liu, *Copyright Law's Theory of the Consumer*, 44 B.C. L. REV. 397, 401 (2003) (dividing the consumer into passive consumer, the consumer as author and active consumer); Niva Elkin Koren, *Making Room for Consumers under the DMCA*, 22 BERKELEY TECH. L.J. 1119, 1138(2007) (indicating that besides the traditional conceptions of consumers

the exploitation of works and thereby enter into public distribution, they do not belong to the category of private use discussed in this book.[49] In other words, private use focuses on those uses that are essentially consumptive.

Moreover, there are actually three stakeholders in a copyright system. In addition to the two camps which are mainly dealt with here – copyright holders and users – there are also intermediaries. The liability of internet service provider (ISP) also remains a very important issue in the digital environment. However, given the fact that intermediaries are less relevant to the private use debate, they will not be discussed in this book.

Various methodologies are employed in this book. Among these institutionalism, economic analysis of law, and comparative study are the main approaches. To begin with this paper attempts to analyze the causes of China's piracy dilemma from an institutionalist perspective.[50] This is done by, for instance, an exploration of two of the limitations particular to an administratively dominant copyright implementation "dual system" based on some concepts of an institutional approach to law, notably Neil Komesar's theory.[51] Pursuant to the capture hypothesis, administrative agencies are likely to become "captured" by the interest groups that they are charged with

---

of copyrighted works, namely the consumer-shopper and consumer-author, there is a third aspect of copyright consumer – the consumer-participant); Julie E. Cohen, *The Place of the User in Copyright Law*, 74 FORDHAM L. REV. 347, 348-49 (2005) (summarizing three kinds of users in literature – the economic user, the "postmodern" user and the romantic user, and indicating a fourth candidate of user – the situated user).

49 Transformative use requires "the use must be productive and must employ the quoted matter in a different manner or for a different purpose from the original". See Pierre N. Leval, *Toward a Fair Use Standard*, 103 HARV. L. REV. 1105 (1990). By comparison, active user's "mini-authorship" does not amount to original authorship, and receives the protection of copyright law as a secondary author. However, active use is not merely passive consumption, but active participation, and places emphasis on autonomy, communication, and creative self-expression. See Liu, *id.* at 415; Koren, *id.* at 1139; Cohen, *id.* at 370 (Cohen proposes the notion of a situated user, who "engages cultural goods and artifacts found within the context of her culture through a variety of activities ranging from consumption to creative play", and appropriates cultural goods for four primary purposes: consumption, communication, self-development, and creative play.).

50 See *infra* section 2.2.

51 See e.g., NEIL KOMESAR, IMPERFECT ALTERNATIVES: CHOOSING INSTITUTIONS IN LAW, ECONOMICS AND PUBLIC POLICY (Chicago and London: The University of Chicago Press 1994) (hereinafter Komesar, Imperfect Alternatives); NEIL KOMESAR, LAW'S LIMITS: THE RULE OF LAW AND THE SUPPLY AND DEMAND OF RIGHTS (Cambridge: Cambridge University Press 2000).

regulating.[52] Concentrated high-stake interests often prevail over dispersed small-stake interests in the decision-making process and this generates a minoritarian bias.[53] This exploratory book is an attempt to elaborate how these theories are embodied in the Chinese context – namely in the multi-headed copyright administration and the agency interests in the copyright legislative process. The aim of this is to highlight the analytic value of these concepts in a very different political and legal system from those in which they are usually studied.

The second approach is based on an economic analysis of law. This will illustrate the limitations of perfect control by TPMs on creativity-enhancing from a dynamic efficiency perspective.[54] This section will review some normative arguments of static efficiency and dynamic efficiency in economics, and highlight the fact that a system operating at short run static efficiency may not be dynamically efficient in the long term. In addition, the cost and benefit analysis of copyright protection reminds us that at low levels of protection the revenue-enhancing effect would dominate, while beyond a certain point higher levels will cause the cost-enhancing effect to dominate.[55] It is therefore of crucial importance to maintain an appropriate level of copyright protection.

Finally, although the geographical locus of this paper is the Chinese digital copyright law, this book draws comparisons with other selected jurisdictions. Concerning the objective of copyright law, the historical divergence of copyright laws in Anglo-US and *droit d' auteur* countries are examined.[56] Based on a comparative investigation of TPMs regulations in the World Intellectual Property Organization (WIPO) Internet Treaties and the DMCA certain recommendations for revisions to the legal treatment of private use in China are put forward.[57] The judicial application of anticircumvention rules and copyright limitations in selected cases from the US and Canada are also explored.[58]

---

52 See Thomas Merrill, *Capture Theory and the Courts: 1967-1983*, 72 CHI.-KENT L. REV. 1039, 1043 (1997).

53 See KOMESAR, IMPERFECT ALTERNATIVES, *supra* note 51, at 76; see also Einer Elhauge, *Does Interest Group Theory Justify More Intrusive Judicial Review?* 101 YALE L. J. 31 (1991).

54 See *infra* section 3.5.2.2.

55 See WILLIAM M. LANDES & RICHARD A. POSNER, THE ECONOMIC STRUCTURE OF INTELLECTUAL PROPERTY LAW 78 (Cambridge: The Belknap Press of Harvard University Press 2003).

56 See *infra* section 3.5.1.1.

57 See *infra* sections 3.2.1 & 3.2.2.

58 See *infra* sections 3.2.3 & 3.2.4.

Regulations and cases from such selected jurisdictions are only provided for the purpose of serving as comparative specimens and are not examined in full detail.

This book is further divided into five chapters. Before turning to the central question of what should be the role of private use in the Chinese digital copyright regime, this paper first illustrates what the current situation is. Chapter I opens with a description of the legal treatment of private use under Chinese copyright law. The first section starts with a narrative description of Chinese copyright law, including the general principles, the protected copyright, and the restrictions on copyright. The following section examines the conventional statutory "free zone" of private use under Chinese copyright law, which is composed of the first sale doctrine and fair use. In the last section, I review the impact of digitization on the legal treatment of private use and argue that the main problem in China is the overprotection of TPMs without due consideration for the legitimate interests of users.

In chapter II, I investigate the causes of such legal treatment and further analyze the underlying reasons for the piracy dilemma in China. In the first section, relying on an investigation of the legislative background of the Chinese digital copyright law, I suggest that the abolishment of the private use "free zone" in digital environment actually serves as an anti-piracy measure. In the subsequent section, I argue that such a treatment is not a reasonable choice because it fails to catch the essential deep-seated causes of the piracy dilemma in China, which are to be found in the law making process, the implementation of the law, adherence to the law by the relevant parties, and in the market.

In the following chapter III, the fierce debates regarding the proper place of private use in the digital environment are addressed. The chapter begins by locating the boundary between copyright exploitation on the one hand and private use and enjoyment on the other and noting the difficulty in separating the two in the digital era. In the following section efforts to harmonize TPMs and private use from both the legislative and judicial perspectives, as well as the tit-for-tat debates on employing TPMs to control private use, are examined. Finally, the objective of Chinese copyright law is employed to evaluate TPMs controls and private use.

Arguably, TPMs are not the sole option for digital content providers to distribute their copyrighted information on the Internet. In the subsequent chapter IV, I compare the "TPMs & contract" model with the other three alternatives – the "levy system" model, the "device-centric" model, and the "cloud" model. I illustrate the pros and cons of each business model concerning private use. In respect of the negative effect, the social conditions required for them to be effective are examined, as well as the potential problems in the Chinese practice which may occur when these conditions are reached.

On the basis of the findings in the above chapters, in the last chapter, the "TPMs & contract" model is used as an example to illustrate the modifications in the law making process, the interpretation and application of the law, and the social infrastructure that will be necessary in China.

Chapter I The Legal Treatment of Private Use under Chinese Copyright Law

The People's Republic of China promulgated its first copyright law and implementing regulations in 1990.[59] Although possessing several limitations, the 1990 Chinese Copyright Law was recognized as a "successful and modern" law that could provide the foundation of a legal framework for copyright protection in China.[60] According to the law, the National Copyright Administration (NCAC) of the People's Republic of China is the competent authority in charge of administrative and enforcement matters related to copyright.[61]

With years of development, China has been gradually improving its system of laws and regulations on copyright, and constantly strengthening their level of enforcement. In order to adapt to international requirements for the protection of intellectual property rights (hereinafter IPRs), China amended the copyright law in 2001 and reissued its implementing regulations.[62] In 2010, the copyright law was revised again but only slightly.[63] Soon, the preparation work on the third amendment of the copyright law began on July 13, 2011.

---

59 Copyright Law of the People's Republic of China was adopted on Sep. 7, 1990 at the 15th Session of the Standing Committee of the 7th National People's Congress [hereinafter the 1990 Chinese Copyright Law]. See also Regulations for the Implementation of Copyright Law of the People's Republic of China. It was promulgated on May 30, 1991, and effective on Jun. 1, 1991 by the National Copyright Administration [hereinafter the 1991 Chinese Copyright Law Implementing Regulations].

60 See Wei Zhi, Cong bo'er'ni gongyue kan zhongguo zhuzuoquan fa zhi xiuding [*An Evaluation on the Revision of the Chinese Copyright Law from the Berne Convention Perspective*], 5 Zhongwai faxue [PEKING UNIVERSITY LAW JOURNAL] 91, 94 (1997).

61 See Article 8 of the 1990 Chinese Copyright Law. For more details of the functions of the NCAC, see *infra* section 1.1.2.3.

62 Copyright Law of the People's Republic of China was amended on Oct. 27, 2001 at the 24th Session of the Standing Committee of the 9th National People's Congress [hereinafter the 2001 Chinese Copyright Law]. See also the new Regulations for the Implementation of Copyright Law of the People's Republic of China. It was promulgated on Aug. 2, 2002, and effective Sep. 15, 2002 by the State Council [hereinafter the 2002 Chinese Copyright Law Implementing Regulations].

63 Copyright Law of the People's Republic of China was amended on Feb. 26, 2010 for the second time at the 13th meeting of the Standing Committee of the 11th National People's Congress [hereinafter the 2010 Chinese Copyright Law]. There were only two provisions being revised. The new Article 4 provides that "Copyright owners, in

In light of its opening-up policy, China has actively acceded to a series of international conventions concerning copyright, such as the Berne Convention for the Protection of Literary and Artistic Works (October 15, 1992), the Universal Copyright Convention (October 30, 1992), and the Geneva Phonograms Convention (April 30, 1993). After accession to the main international conventions and treaties on copyright and other intellectual property, China acceded to the WTO on December 11, 2001, which indicated that its "[i]ntegration into the international intellectual property community was completed".[64]

This section will examine the legal treatment of private use under Chinese copyright law, and especially the impact of digitization. For ease of understanding, this section begins with a brief introduction to the legal framework of Chinese copyright law.

1.1 Legal Framework of Chinese Copyright Law

This section introduces some fundamental issues of Chinese copyright law, including the general principles of the law, the categories of copyright, the standards for copyright infringement, and the "dual system" approach to copyright protection. Also discussed are the copyright limitation system and especially the doctrine of fair use within Chinese copyright law.

1.1.1 General Principles

The Copyright Law of the People's Republic of China was enacted in accordance with the Chinese Constitution. It was promulgated for the purposes of protecting the copyright of authors in their literary, artistic and scientific works and rights related to copyright, of encouraging the creation and dissemination of works which would contribute to the construction of socialist spiritual and material civilization, and of promoting the development and flourishing of socialist culture and sciences.[65]

---

exercising their copyright, shall not violate the Constitution or laws or infringe upon the public interests. The publication and dissemination of works shall be subject to the administration and supervision of the state." The newly added Article 26 provides that "In case of pledge of copyright, the pledger and the pledgee shall go through registration of the pledge with the copyright administration under the State Council."

64 INTELLECTUAL PROPERTY LAW IN CHINA, Introduction, xiii (Peter Ganea, Thomas Pattloch & Christopher Heath eds., Hague: Kluwer Law International 2005).

65 See Article 1 of the 2010 Chinese Copyright Law.

The works of all Chinese citizens, legal entities or other organizations, regardless of whether they are published or not, are entitled to copyright protection. The works of foreigners or stateless persons published for the first time and within the territory of China also enjoy copyright. The works of foreigners or stateless persons published outside China are entitled copyright protection pursuant to bilateral agreements between the country to which the author belongs or in which the author permanently resides and China, or international treaties to which both countries are parties.[66]

The term "works" used in the Chinese Copyright Law refers to original intellectual creations in the literary, artistic and scientific domain, insofar as they are capable of being reproduced in a certain tangible form.[67] The Chinese copyright law defines "works" to include the following types of publications in such fields as literature, art, natural science, social science, engineering technology and the like: (1) written literary works; (2) orally delivered works; (3) musical, dramatic, quyi [folk ballads and story-telling, etc.], choreographic and acrobatic art works; (4) works of fine art and architecture; (5) photographic works; (6) cinematographic works and works created in a way similar to cinematography; (7) drawings of engineering designs and product designs, maps, sketches and other graphic works as well as model works; (8) computer software;[68] (9) other works as provided in laws and administrative regulations.[69]

66 See Article 2 of the 2010 Chinese Copyright Law. Notably, Article 2 also provides that "[w]orks of foreigners or stateless persons shall receive protection in accordance with this Law where the relevant country has not signed an agreement or joined an international treaty with China but where the works have first been published in a member state of the international treaty to which China is a participant, or have been published simultaneously in member and non-member states of the said treaty." Article 7 of the 2002 Chinese Copyright Law Implementing Regulations provides that works of foreigners or stateless persons published for the first time and within the territory of China shall enjoy copyright from the day the work is published for the first time. Article 8 of the 2002 Chinese Copyright Law Implementing Regulations provides that works of foreigners and stateless persons first published outside of China shall be deemed to be published in China at the same time if publish in China within 30 days.

67 See Article 2 of the 2002 Chinese Copyright Law Implementing Regulations.

68 Besides the 2010 Chinese Copyright Law, a number of administrative regulations have been promulgated specifically addressing computer software. See Regulations on Computer Software Protection (Order No. 339 of the State Council, promulgated on Dec. 20, 2001, and effective from Jan. 1, 2002); Measures for Computer Software Copyright Registration (issued on Apr. 6, 1992 by the Ministry of Electronics Industries of the People's Republic of China, and amended on May 26, 2000).

69 See Article 3 of the 2010 Chinese Copyright Law. In order to protect subject matter provided in Article 2 of the Berne Convention, the 2010 Chinese Copyright Law includes architectural works and three-dimensional models in the category of protected

Copyright owners, in exercising their copyright, must not violate the Constitution or laws or infringe upon the public interests. The publication and dissemination of works are subject to the administration and supervision of the state.[70]

Notably, copyright protection is not available to laws, regulations, government-issued resolutions, decisions or orders, other documents of legislative, administrative or judicial nature, or their official translations. The Chinese Copyright Law also excludes news on current affairs, calendars, numerical tables, forms of general use, and formulas from the category of protected works.[71]

1.1.2 Protected Copyright

Copyright holders under the Chinese copyright law include authors and other citizens, legal entities and organizations enjoying copyright in accordance with the Law.[72] In this sense, both natural and artificial persons could possess copyright. In addition, not only can human creators of works own copyrights in the first instance, but transferees may also receive ownership through property transactions, or succession after the author's death.

1.1.2.1 Moral Rights and Economic Rights

There are two categories of copyright under the Chinese copyright law: moral rights and economic rights.[73] Moral rights protect an author's non-pecuniary or

---

works. The addition of acrobatic works in the list of protected works can be regarded as an attempt to protect the rich and long-standing Chinese arts culture. Article 6 of the 2010 Chinese Copyright Law provides that regulations for the protection of copyright in expressions of folklore shall be separately established by the State Council.

70 See Article 4 of the 2010 Chinese Copyright Law.

71 See *id.* Article 5.

72 See *id.* Article 9.

73 See Article 10 of the 2010 Chinese Copyright Law. The recognition of dual copyright in Chinese copyright law is in accordance with international treaties. E.g., Article 27 (2) of the 1948 Universal Declaration of Human Rights provides: "everyone has the right to the production of the moral and material interests resulting from any scientific, literary or artistic production of which he is the author". Article 15 (1) of the International Covenant on Economic, Social and Cultural Rights 1966 provides that: "the States Parties to the present covenant recognize the right of everyone…to benefit from the protection of the moral and material interests resulting from any scientific, literary or

non-economic interests.[74] An author has a right to claim authorship of the work, and to object to any distortion, mutilation or other modification of, or other derogatory treatment of the work prejudicial to the author's honor or reputation.[75]

Moral rights under the Chinese copyright law include the right of publication,[76] the right of authorship,[77] the right of alteration,[78] and the right of integrity.[79] There is no limitation on the duration of an author's moral rights, except the right of publication.[80] On account of the non-transferable character of moral rights, an author may continue to exercise them in respect of a work even after the transfer of the copyright ownership, or the expiry of economic rights. Moreover, moral rights are protected even after the death of the author.

In contrast with the non-transferable and ever-lasting moral rights, economic rights arising from copyright ownership in protected works can be transferred by the author to other people or organizations, and their protection expires after a certain period.[81] The scope of economic rights has steadily expanded as the development of information transmission technologies. Economic rights in the 1990 Chinese Copyright Law were vague and limited both in content and form. Only the right of exploitation and the right to remuneration were mentioned.[82] In the 2001 Chinese Copyright Law, different types of economic rights were

artistic production of which he is the author." See also Article 6 *bis* and Article 14 *ter* of the Berne Convention.

74 See Jane C. Ginsburg, *Moral Rights in a Common Law System*, 1 ENT. L. REV. 121 (1990).

75 See Article 6 *bis* of the Berne Convention.

76 See Article 10 (1) of the 2010 Chinese Copyright Law (providing the right to decide whether to make a work available to the public).

77 See *id.* Article 10 (2) (providing the right to claim authorship and to have the author's name mentioned in connection with the work).

78 See *id.* Article 10 (3) (providing the right to alter or authorize others to alter one's work).

79 See *id.* Article 10 (4) (providing the right to protect one's work against distortion and mutilation).

80 See *id.* Article 20 ("The rights of authorship, alteration and integrity of an author shall be unlimited in time.").

81 See *id.* Article 21 (providing that the term of protecting economic rights shall be the lifetime of the author and fifty years after his death, expiring on Dec. 31 of the fiftieth year after his death.).

82 See Article 10 (5) of the 1990 Chinese Copyright Law ("[T]he right of exploiting one's work by means of reproduction, performance, broadcasting, exhibition, distribution, making cinematographic, television or video production, adaptation, translation, annotation, compilation and the like, and the right of authorizing others to exploit one's work by the above-mentioned means, and of receiving remuneration therefor.").

clarified and further expanded to cover the right of reproduction,[83] the right of distribution,[84] the right of lease,[85] the right of exhibition,[86] the right of performance,[87] the right of projection,[88] the right of broadcasting,[89] the right of information network transmission,[90] the right of production,[91] the right of adaptation,[92] the right of translation,[93] the right of compilation,[94] and other rights which are enjoyed by the copyright owners.[95]

1.1.2.2 Standards for Copyright Infringement

The 2010 Chinese Copyright Law explicitly enumerates the types of copyright infringements. Some copyright infringements only bear civil liability, while

83 See Article 10 (5) of the 2001 Chinese Copyright Law ("[T]he right to produce one or more copies of the work by means of printing, Xeroxing, rubbing, sound recording, video recording, duplicating, or re-shooting, etc".).

84 See *id.* Article 10 (6) ("[T]he right to provide the public with original copies or reproduced copies of works by means of selling or donating".).

85 See *id.* Article 10 (7) ("[T]he right to non-gratuitously permit others to temporarily exploit a cinematographic work, a work created in a way similar to cinematography or computer software, unless the computer software is not the main object under the lease".).

86 See *id.* Article 10 (8) ("[T]he right to publicly display the original copies or reproduced copies of works of fine art and cinematographic works".).

87 See *id.* Article 10 (9) ("[T]he right to publicly perform works, and to publicly transmit the performance of works by various means".).

88 See *id.* Article 10 (10) ("[T]he right to make, by such technical equipment as projector, episcope, etc., the works of fine art, photographic works, cinematographic works and works created in a way similar to cinematography, etc. reappear publicly".).

89 See *id.* Article 10 (11) ("[T]he right to publicly broadcast or disseminate works by wireless means, to disseminate broadcast works to the public by wired dissemination or rebroadcast, and to disseminate broadcast works to the public by audio amplifier or other similar instruments for transmission of signs, sounds or images".).

90 See *id.* Article 10 (12) ("[T]he right to provide the public with works by wired or wireless means, so as to make the public able to respectively obtain the works at the individually selected time and place".).

91 See *id.* Article 10 (13) ("[T]he right to fix works on the carrier by cinematography or in a way similar to cinematography".).

92 See *id.* Article 10 (14) ("[T]he right to modify a work for the purpose of creating a new work of original creation".).

93 See *id.* Article 10 (15) ("[T]he right to transform the language of a work into another language".).

94 See *id.* Article 10 (16) ("[T]he right to choose or edit some works or fragments of works so as to form a new work".).

95 See *id.* Article 10 (17).

some infringements might simultaneously bear administrative liability when damage is caused to the public interest, or criminal liability when the action constitute a crime.

Article 47 Anyone who commits any of the following acts of infringement shall bear the civil liability for such remedies as ceasing the infringing act, eliminating the effects of the act, making a public apology or paying compensation for damages, depending on the circumstances:

(1) publishing a work without the permission from the copyright owner;

(2) publishing a work of joint authorship as a work created solely by oneself, without the permission from the other co-authors;

(3) having one's name mentioned in connection with a work created by another, in order to seek personal fame and gain, where one has not taken part in the creation of the work;

(4) distorting a work created by another;

(5) plagiarizing the works of others;

(6) exploiting a work by means of exhibition, making cinematographic productions or a means similar to making cinematographic productions, or by means of adaptation, translation, annotation, etc. without the permission from the copyright owner, unless otherwise provided in this Law;

(7) exploiting a work of another without paying the remuneration;

(8) without the permission from the copyright owner or obligee related to the copyright of a cinematographic work or a work created in a way similar to cinematography, computer software, sound recordings or video recordings, leasing one's work or sound recordings or video recordings, except where otherwise provided in this Law;

(9) without the permission from a publisher, exploiting the format design of one's published book or periodical;

(10) without the permission from the performer, broadcasting or publicly transmitting his live performance or recording his performance;

(11) committing other acts of infringement upon copyright and upon other rights related to copyright.

Article 48 Anyone who commits any of the following acts of infringement shall bear the civil liability for such remedies as ceasing the infringements, eliminating the effects of the act, making a public apology or paying compensation for damages, depending on the circumstances; where he damages public interests at the same time, the copyright administration department may order him to cease the act of tort, may confiscate his illegal gains, confiscate and destroy the reproductions of infringement, and impose a fine on him; if the case is serious, the copyright administration department may also confiscate the materials, instruments and equipment, etc. mainly used to make the reproductions of infringement; where his act has constituted a crime, he shall be investigated for criminal liabilities in accordance with the law:

(1) without the permission from the copyright owner, reproducing, distributing, performing, projecting, broadcasting, compiling, disseminating to the public through information network one's works, except where otherwise provided in this Law;

(2) publishing a book where the exclusive right of publication belongs to another;

(3) without the permission from a performer, reproducing, distributing the sound recordings or video recordings of his performance, or disseminating his performance to the public through information network, except where otherwise provided in this Law;

(4) without the permission from a producer of sound recordings and video recordings, reproducing, distributing, disseminating to the public through information network the sound recordings or video recordings produced by him, except where otherwise provided in this Law;

(5) without the permission, broadcasting or reproducing the radio or television, except where otherwise provided in this Law;

(6) without the permission from the copyright owner or obligee related to the copyright, intentionally avoiding or destroying the technical measures taken by the obligee on his works, sound recordings or video recordings, etc. to protect the copyright or the rights related to the copyright, except where otherwise provided in laws or administrative regulations;

(7) without the permission from the copyright owner or obligee related to the copyright, intentionally deleting or altering the electronic information on the management of the rights on the works, sound recordings or video recordings, except where otherwise provided in laws or administrative regulations;

(8) producing or selling a work where the signature of another is counterfeited.

In order to repair the potential limitation of the enumerated list of copyright infringements, Article 47 (11) contains a flexible regulation referring other infringements upon copyright and upon other rights related to copyright. This elastic regulation is usually interpreted to include other types of primary infringement of copyright absent from the enumerated list, and secondary infringement of copyright.[96] The division between primary infringement and secondary infringement of copyright is recognized in the theoretical field in Chinese copyright.[97] However, on account of the lack of explicit regulation of secondary infringement of copyright in the statute, courts have customarily

96 See Wang Chunyan, Lun zhuzuoquan qinquan xingwei de tedian jiqi rending [*Characteristics and Determination of Copyright Infringement*], 2 Zhongwai faxue [PEKING UNIVERSITY LAW JOURNAL] 39, 42 (1996).

97 See Liu Chunmao ed., Zhishi chanquan yuanli [FOUNDATIONS OF IP] 322 (Beijing: Zhishi chanquan chubanshe [IP Publishing Company] 2002).

invoked the provision of contributory infringement in the General Principles of Chinese Civil Law.[98]

1.1.2.3 "Dual System" Approach to Copyright Protection

As Roscoe Pound has demonstrated, the given geographical environment, the customs and habits of the people and general economic and historical conditions demand that legal regulation and governance should adapt to the local conditions.[99] Therefore, it is of crucial importance in the first instance to scrutinize "how the past influences the present and the future".[100] The institutional choice of parallel dual copyright protection systems is determined by the distinct Chinese national conditions.

In the initial stages of establishing a copyright regime, given the exotic character of the copyright enterprise and the necessity of starting from scratch, China adopted the so-called copyright "dual system". This was intended to exert the advantages of two parallel approaches – the judicial process as well as administrative protection – and to establish the final ruling of justice. This was felt to be the best option given the insufficiency of the transition period[101] as well as the lack of necessary preconditions for effective implementation.[102]

---

98 See General Principles of the Civil Law of the People's Republic of China (Adopted at the Fourth Session of the Sixth National People's Congress, and promulgated by Order No. 37 of the president of the People's Republic of China on Apr. 12, 1986, and effective on Jan. 1, 1987). Article 130 provides that "[i]f two or more persons jointly infringe upon another person's rights and cause one damages, they shall bear joint liability."

99 Roscoe Pound, Fali xue [JURISPRUDENCE] 2 (vol.1, Yu lüxue Trans., Beijing: Falü chubanshe [Law Press China] 2007).

100 DOUGLASS NORTH, INSTITUTIONS, INSTITUTIONAL CHANGE AND ECONOMIC PERFORMANCE 3(1990).

101 Globalization and trade integration deprived China of the necessary transitional process of copyright protection from low level into high level. For instance the transition period for China only lasted about 10 years, while this period was sustained from a period of several decades to hundreds of years in Western countries.

102 To achieve the policy target of the copyright system a series of preconditions, including substantial establishments in the area of economy, technology, culture and education, as well as the social environment and public policy scheme are relied on. Meanwhile, the court universally lacked experience to deal with new types of IPR cases which were dramatically distinctive from traditional civil, administrative and criminal cases, and at the same time lacked experienced professionals who were familiar with IPRs.

Within the "dual system", administrative protection is the notable characteristic possessing significance.[103] In addition to granting private relief to the right holder it is both reasonable and necessary for public power to intervene in the protection of copyright. This flows from copyright's position as an important private right which simultaneously involves a public right characteristic that is closely related to economic and social development.[104] Moreover, in light of public interest theory, administrative agencies would be "better able to act in the public interest".[105]

The "dual system" has lasted more than 20 years, and played a significant role in the establishment and implementation of the copyright regime. The former secretary general of the WIPO, Arpad Bogsch, praised China because, with regard to the protection of IPRs, it has traversed the path within 25 years for which the industrialized nations needed more than 100 years.[106] The dramatic achievement of protecting copyright and other IPRs has proved the

---

103 All previous texts of three main IPR legislations have regulated judicial and administrative protection, especially in patent law and trademark law, which include abundant clauses regulating administrative protection in every detail. Comparatively, the 2010 Chinese Copyright Law includes fewer clauses relating to administrative protection, i.e. Articles 7, 28, 48, 55 & 56.

104 See Feng Xiaoqing & Liu Shuhua, Shi lun zhishi chanquan de siquan shuxing jiqi gongquanhua quxiang *[On the Private Right Nature of Intellectual Property and Its Public Trend*], 1 Zhongguo faxue [CHINESE LEGAL SCIENCE] 61 (2004); Li Yongming & Lü Yilin, Lun zhishi chanquan zhi gongquan xingzhi –dui "zhishi chanquan shuyu siquan" de buchong [*Public Authority of Intellectual Property*], 4 Zhejiang daxue xuebao [JOURNAL OF ZHEJIANG UNIVERSITY] 60 (2004).

105 Samuel Huntington, *The Marasmus of the ICC: The Commission, the Railroads, and the Public Interest*, 61 YALE L.J. 467, 508-09 (1952); see also Robert Rabin, *Federal Regulation in Historical Perspective*, 38 STAN. L. REV. 1189, 1272-78 (1986).

106 See Guojia zhishi chanquan ju [STATE INTELLECTUAL PROPERTY OFFICE (SIPO)], 1994 nian zhongguo zhishi chanquan baohu zhuangkuang (baipi shu) [CHINESE IPRS PROTECTION STATUS IN 1994]. However, on account of the lack of the necessary transition period and the corresponding internalization conditions, high level transplanted legislations do not operate and implement well in practice. According to statistics obtained by quantitative analysis, early in 1993 the protection level, without considerating actual implementation, in China (score 3.190) had already exceeded part of that of developed countries; in 2001, the score increased to 4.19 which exceeded almost all developed countries and developing countries, and was only slightly inferior to that of the US (score 4.52). When amended by the implementation index, however, the protection level score decreased to 2.742. See Han Yuxiong & Li Huaizu, Guanyu zhongguo zhishi chanquan baohu shuiping de dingliang fenxi [*Quantitative Analysis on Chinese IPRs Protection Level*], 3 Kexuexue yanjiu [STUDIES IN SCIENCE OF SCIENCE] 377 (2005).

rationality of the "dual system" which makes use of the advantages and mutual complementarities of alternative arenas of protection.

In contrast with judicial protection, the administrative approach possesses distinct institutional advantages. First, a centralized source of governmental authority which combines all governmental powers under one convenient roof secures coordinated solutions to IPR disputes throughout its jurisdiction. Notably special, well-planned and focused actions are accomplished when dealing with cases of repeated, organized IPR infringements, large-scale counterfeiting and piracy. Individual case based verdicts, together with the non-recognition of precedent in China, at the same time restrict the influence of judicial decisions to a very limited scope.

Second, compared with the formalized judicial process and the high costs of access to the courts, administrative protection enjoys the advantage of high efficiency and low costs, which is in accordance with the general principle of the TRIPS that protecting IPRs should be achieved with minimized cost.[107]

Third, in contrast with passive judicial protection following the principle of "no trial without complaint", administrative agencies can actively intervene in and solve disputes before they are upgraded into lawsuits. Moreover, the staff of administrative agencies is expected to have specialized information and systematic knowledge – in other words, the expertise – to solve the complex problems that arise in this rapidly changing field.[108]

Finally, given that the dual system provides for judicial review of administrative settlements, it does not exclude the opportunity for an alternative judicial process,[109] but rather provides an extra protecting shield for right holders. If the parties refuse to accept the administrative decision, they can still go to the courts as a last resort. In this sense, administrative protection can be understood as a preliminary procedure which could provide complementary support to the judicial process and jointly maintain the order of copyright protection.

1.1.3 Restrictions on Copyright

Copyright is a body of law which accommodates the competing interests of various groups in our society. From its statutory beginnings in early-eighteenth-century England, copyright law has been the product of a precarious attempt to balance the rights of the creators and those of their users, both present and

107 See Articles 41-61 of the TRIPS.

108 See Merrill, *supra* note 52, at 1049.

109 Except for in some special circumstances where the administration ruling is final.

future.[110] The balancing of interests is a basic requirement and a general object of a copyright regime, a fact that has been embodied in both international treaties and domestic laws. In this section, I intend to introduce the Chinese copyright limitation system, and especially the fair use doctrine, as a built-in balancing tool.

1.1.3.1 Chinese Copyright Limitation System in General

The overall design of the Chinese copyright regime strikes a delicate balance between the interests of copyright holders in the control and exploitation of their copyrighted materials on the one hand, and society's public interests in the free flow of information, and dissemination of works which would contribute to the development and flourishing of socialist culture and sciences on the other.[111]

With the specific intent of encouraging the production and distribution of new works for the public, Chinese copyright law provides incentive for creators by granting them a series of exclusive rights to reproduce and distribute their work.[112] However, these rights are not absolute, as is the tangible property right, but are subject to various limitations.

Within the Chinese copyright regime, a number of inherent limits are designed to promote the creation and dissemination of new works as well as to ensure the preservation of a vigorous public domain.[113] Such inherent limits include the limited duration of the copyright protection,[114] the requirement of originality,[115] the dichotomy of idea and expression,[116] and the first sale doctrine.[117]

---

110 See PATTERSON & LINDBERG, *supra* note 9, at 2.

111 See Article 1 of the 2010 Chinese Copyright Law: "This Law is enacted, in accordance with the Constitution for the purposes of protecting the copyright of authors in their literary, artistic and scientific works and rights related to copyright, of encouraging the creation and dissemination of works which would contribute to the construction of socialist spiritual and material civilization, and of promoting the development and flourishing of socialist culture and sciences."

112 See *id.* Article 10.

113 See Jessica Litman, *The Public Domain*, 39 EMORY L.J. 965, 975 (1990).

114 See Article 21 of the 2010 Chinese Copyright Law.

115 See Article 2 of the 2002 Chinese Copyright Law Implementing Regulations.

116 Article 6 of the Regulations on the Protection of Computer Software (promulgated on Dec. 20, 2001 by the State Council, No.339, effective on Jan. 1, 2002) provides the dichotomy of idea and expression ("The protection provided to software copyright under these regulations cannot be expanded to encompass the ideas, processing courses, operation methods or mathematical concepts, etc. used in the development of computer software."). Although the 2001 Chinese Copyright Law fails to contain the dichotomy

Apart from the Chinese copyright regime's inherent limits, a balance of interests between encouraging the author's enthusiasm of further creation and the user's benefit of knowledge sharing and dissemination is achieved through the recognition of copyright limitations. There are several words used to refer to the derogations from copyright, such as "limitations", "restrictions", "permissions", "defence", "permitted acts",[118] and "exceptions".[119] Notably, a lot of scholars prefer to choose "limitations" for the reason that it is more neutral and broader than other words, and thus more capable of taking account of the characteristics of a different copyright regime.[120] Moreover, the choice between labeling fair use as an "exception" or "limitation" is a pedagogical one reaching into the very heart of copyright policy.[121]

There are three main categories of copyright limitations[122] in Chinese copyright law: fair use exemptions,[123] statutory licenses,[124] and mandatory

---

of idea and expression explicitly, this principle is widely recognized in the copyright judicial practice. For example, in 1990 in *Li Shuxian, Wang Qingxiang v. Jia Yinghua* Beijing shi xicheng qu renmin fayuan [Beijing Xicheng district People's Court] made the verdict in light of the dichotomy of idea and expression. In 2006, in *Wang Tiancheng v. Zhou Yezhong, Dai jitao, ren min chu ban she*, Beijing shi di'er zhongji renmin fayuan [The Beijing Second Intermediate People's Court] definitely indicated in the decision that "copyright protects the expression of idea, rather than the idea itself". See the civil verdicts of The Beijing Second Intermediate People's Court, No. 06122 (2006), available at http://www.dffy.com/sifashijian/ws/200607/20060721205003-3.htm (last visited Jun. 10, 2010). See also Zhishi chanquan anli pingxi [IP CASE REVIEW] 2-7 (Zheng Chengsi ed., Beijing: Falü chubanshe [Law Press China] 1994).

117 See *infra* sections 1.2.1 and 1.3.2.

118 See Chapter III of Part 1 the UK Copyright Designs and Patent Act 1988.

119 E.g., the Berne Convention and the EU InfoSoc Directive prefer the term of "exceptions".

120 See e.g., LUCIE M.C.R. GUIBAULT, COPYRIGHT LIMITATIONS AND CONTRACTS: AN ANALYSIS OF THE CONTRACTUAL OVERRIDABILITY OF LIMITATIONS ON COPYRIGHT 16 (The Hague; London: Kluwer Law International 2002).

121 As Ruth Okediji questions: "Is fair use merely an adjunct to copyright law serving perfunctory purposes of a narrow nature and, thus, merely an 'exception' or is it a 'limitation', an integral part of what domestic copyright law purports to be, namely an instrument of public policy designed first to allocate resources and then to redistribute those resources to facilitate the purposeful end of progress in the arts?" See Ruth Okediji, *Toward an International Fair Use Doctrine*, 39 COLUM. J. TRANSNAT'L L. 75, 113-14 (2000).

122 This classification is in accordance with Lucie Guibault. See GUIBAULT, *supra* note 120, at 20-26 (stating that the essential difference between continental European's "naturalist" conception and the American notion of a "utilitarian" copyright system is also visible in the way the limitations are expressed in legislation. No matter whether formulated in open or closed terms, copyright limitations can generally be classified

collective administration of author's rights.[125] Notably, although the regulations on compulsory licenses of the Berne Convention could be applied in its member state China,[126] the current domestic Chinese copyright law does not contain any regulation of compulsory licenses.[127]

Among various categories of copyright limitations, the most common form recognized in favor of users consists of a full exemption from the scope of the copyright owner's exclusive rights. The fair use exemption provides a statutory permission for the user to make use of a copyrighted work and even the use of copyright.[128] On account of the fact that they carve out certain uses from the scope of the copyright holder's exclusive right to prohibit, and deprive the holder of the economic reward or compensation for their intellectual labor, exemptions are recognized as "the most serous encroachment upon the copyright owner's exclusive rights".[129]

Having introduced the system of Chinese copyright limitations, the following section explores the meaning and categories of "fair use" under the Chinese copyright law.

1.1.3.2 "Fair Use" under the Chinese Copyright Law

In respect of the fair use concept, there are mainly three categories of narratives in the Chinese copyright field. The first type highlights the essential attribute or the connotation of fair use, namely that "without permission from and without payment of remuneration to the copyright holder". For example, "fair use is the use of copyrighted works which does not need the permission from and payment of remuneration to the right holder, and does not constitute copyright

---

into four main categories: exemptions, statutory licenses, compulsory licenses and mandatory collective administration of author's rights.).

123 See Article 22 of the 2010 Chinese Copyright Law.

124 See *id.* Articles 23, 33(2), 40(3), 43(2) and 44.

125 See *id.* Article 8.

126 See Berne Convention, Article 11 *bis* (2), and Articles 2, 3 and 4 of its Appendix.

127 Some Chinese scholars argue to establish the compulsory license system in the Chinese copyright law. See e.g., Yang Hongjun, Banquan qiangzhi xuke zhidu lun [*The Compulsory License System of Copyright*], 4 Zhishi chanquan [INTELLECTUAL PROPERTY] 30 (2008).

128 See Article 22 of the 2010 Chinese Copyright Law (providing that in the enumerated 12 cases, a work may be used without permission from, and without payment of remuneration to, the copyright owner).

129 See GUIBAULT, *supra* note 120, at 21.

infringement".[130] Or "fair use is the use of works under legal circumstances, which may not get permission from the author, and even does not need to pay remuneration".[131]

The second kind of narrative emphasizes the extension of all objects covered by fair use for a "legitimate purpose" or to an "equitable or reasonable scope". Such as "fair use is the use that one can moderately quote or reproduce other's copyrighted works for the purpose of academic research, literature criticism or other legitimate reasons".[132]

The third way of defining fair use is to combine the connotation of "authorization-free and remuneration-free use" and the extension of "legitimate purpose" together. For example, "fair use is the use of another person's published works without permission from or payment to the copyright holder, as long as the use is for the purpose of private study, research or self-entertainment or for the project of education, scientific research, religion or philanthropy".[133]

One influencing and widely accepted definition of fair use put forward by Wu Handong adopts the third kind of narrative. He argues that fair use is "under legal circumstances, the legitimate act of using another person's copyrighted works for proper purposes, without permission from or payment to the copyright holder".[134]

In substance, fair use is a factual act which leads to certain legal effects according to its actual consequences.[135] As the action of a human being, the factual act of fair use is different from a natural action involving the human body. Moreover, in the field of legitimate actions, the factual act is not merely expressive but induces actual consequences. Thus, it has essential differences from the legal act and the quasi-legal act. Furthermore, because of its legitimate quality, the factual act is also different from the tort, which belongs to the

130 Zheng Chengsi, Zhishi chanquan fa tonglun [GENERAL SURVEY ON INTELLECTUAL PROPERTY LAW] 124 (Beijing: Falü chubanshe [Law Press China] 1986).

131 Duan Ruilin, Zhishi chanquan fa gailun [AN INTRODUCTION TO INTELLECTUAL PROPERTY LAW] 289 (Beijing: Guangming ribao chubanshe [Guangming Daily Publishing House] 1988).

132 Yang Chongsen, Zhuzuoquan zhi baohu [PROTECTION OF COPYRIGHT] 102 (Taiwan: Taiwan zhengzhong shuju [Taiwan Zhengzhong Publishing House] 1977).

133 Shen Rengan, Guanyu dui zhuzuoquan de xianzhi [*On the Restrictions of Copyright*], in Zhonghua Renmin Gongheguo zhuzuoquan fa jiangxi [ANALYSIS ON THE CHINESE COPYRIGHT LAW] 199 (Beijing: Zhongguo guoji guangbo chubanshe [China International Broadcasting Press] 1991).

134 Wu Handong, Zhuzuoquan heli shiyong zhidu yanjiu [RESEARCH ON THE FAIR USE OF COPYRIGHT] 144 (Beijing: Zhongguo zhengfa daxue chubanshe [China University of Political Science and Law Press] 1996).

135 See *id.*

category of illegitimate acts.[136] This understanding of fair use as a factual act in China is quite different from the doctrine of fair use as a defense to a claim for copyright infringement in American copyright law.[137]

As a limitation of the copyright holder's exclusive right or a voluntary sacrifice of the right holder's economic interests for the sake of the pubic interest, fair use is only permitted in specific legal circumstances. Since the introduction of its first copyright law, China has adopted an exhaustive list of fair use. 12 circumstances of fair use were explicitly enumerated in the 1990 Chinese Copyright Law.[138] Some of them were further elaborated and amended in the 2001 Chinese Copyright Law[139] and this was followed in the 2010 Chinese Copyright Law.[140] The exhaustive list of fair use under Chinese copyright law includes the activities as follows:

> Article 22 In the following cases, a work may be exploited without the permission from, and without payment of remuneration to, the copyright owner, provided that the name of the author and the title of the work are mentioned and the other rights enjoyed by the copyright owner by virtue of this Law are not infringed upon:
>
> (1) use of a published work for the purposes of the user's own private study, research or self-entertainment;
>
> (2) appropriate quotation from a published work in one's own work for the purposes of introduction of, or comment on, a work, or demonstration of a point;
>
> (3) inevitable reappearance or citation of a published work in newspapers, periodicals, radio stations, television stations or other media for the purpose of reporting current events;
>
> (4) reprinting by newspapers or periodicals or other media, or rebroadcasting by radio stations or television stations or other media, of the current event articles on the issues of politics, economy and religion, which have been published by other

136 Chang Peng'ao, Shishi xingwei de jichu guifan [*Basic Normative Issues of Factual Acts*], in 1 Faxue yanjiu [CHINESE JOURNAL OF LAW] 48 (2010).

137 The most commonly asserted defense to a copyright infringement claim is that the alleged action is a fair use of the copyrighted works. The US Copyright Act refers to the term "fair use": the fair use of a copyrighted work, including such use by reproduction in copies or phonorecords or by any other means specified by that section, for purposes such as criticism, comment, news reporting, teaching (including multiple copies for classroom use), scholarship, or research, is not an infringement of copyright. See 17 U.S.C. § 107.

138 See Article 22 of the 1990 Chinese Copyright Law.

139 See Article 22 of the 2001 Chinese Copyright Law. For example, in the regulation of free of charge performance, the restriction of fair use by the State organ must be within the reasonable scope for the purpose of fulfilling its official duties, and the limitation of reprinting published works by newspapers or periodicals or other media. However, there is no change regarding private use.

140 See Article 22 of the 2010 Chinese Copyright Law.

newspapers, periodicals, radio stations or television stations or other media, except where the author has declared that publication or broadcasting is not permitted;

(5) publication in newspapers or periodicals or other media, or broadcasting by radio stations or television stations or other media, of a speech delivered at a public assembly, except where the author has declared that publication or broadcasting is not permitted;

(6) translation or reproduction, in a small quality of copies, of a published work for use by teachers or scientific researchers in classroom teaching or scientific research, provided that the translation or reproduction is not published or distributed;

(7) use of a published work by a State organ within the reasonable scope for the purpose of fulfilling its official duties;

(8) reproduction of a work in its collections by a library, archive, memorial hall, museum, art gallery or similar institution, for the purpose of the display or preservation of a copy of the work;

(9) free of charge performance of a published work, that is, with respect to the performance, neither fees are charged from the public nor the remuneration is paid to the performers;

(10) copying, drawing, photographing, or video recording of an artistic work located or on display in an outdoor public place;

(11) translation of a work published by a Chinese citizen, legal entity or organization, which is created in the Han language (Chinese), into a minority nationality language for publication and distribution within the country;

(12) translation of a published work into Braille and publication of the work so translated.

In contrast with an openly interpreted fair use defense in the US, China and most continental European countries have conventionally adopted strictly defined and closed legislation which specifically enumerates certain circumstances that constitute fair use. By enumerating all of the fair use circumstances, legislators intend to make the abstract fair use principle clearer and easier for the public to follow and for judges to apply. On the other hand, however, this explicitly limited enumeration has been widely criticized for its lack of flexibility and inability to adapt to new situations engendered by the development of technology.[141]

141 See Lu Haijun, Heli shiyong yiban tiaokuan de caixiang yu zhengming: heli shiyong zhidu lifa moshi tantao [*General Clause of Fair Use: Research on the Legislative Model of Fair Use*], 2 Zhengfa Luncong [JOURNAL OF POLITICAL SCIENCE AND LAW] 42, 46-47 (2007); Yu Yu & Ji Xiaoxin, Woguo zhuzuoquan heli shiyong panduan biaozhun de fansi yu chonggou [*The Reflection and Reconstruction of the Assessment Criterion for the Fair Use of Copyright in China*], 3 Faxue luntan [LEGAL FORUM] 90, 94 (2007).

1.2 Statutory "Free Zone" of Private Use under Chinese Copyright Law

Chinese copyright law has rather well-developed theories and a system for protecting the authors of works,[142] but it has no similarly well-developed conception of the user. There is no explicit term or concrete definition of "private use" in Chinese copyright legislation.

Arguably, this ignoring of the user as a concept is not a problem that exists solely in China. None of the provisions of the seven copyright-related directives adopted by the European legislature since 1999, for instance, even mention the word "consumer".[143] Even now, the user remains almost completely invisible in the law of European copyright.[144]

The same can be said of current US copyright law. Of course, it would not be true to say that the law totally disregards the interests of users. The existenc of rules on exceptions to copyright, for example, indicates that attention is given to the public including users.[145] After all, the overall purpose of the US

142 The Chinese copyright scheme has a relative well-developed theory about the connotation of the author, which is not limited to the romantic individual, but also includes the collective and corporate authorship. The copyright system grants the author various exclusive rights, and provides economic incentives to engage in creative activity. For more details of the protected copyright, including various moral rights and economic rights, see *supra* section 1.1.2.

143 See Council Directive 91/250 on the Legal Protection of Computer Programs, 1991 O.J. (L 122) 42 (EC); Council Directive 92/100 on Rental and Lending Rights and Certain Rights Related to Copyright in the Field of Intellectual Property, 1992 O.J. (L 346) 61 (EC); Council Directive 93/83 on the Co-Ordination of Certain Rules Concerning Copyright and Rights Related to Copyright Applicable to Satellite Broadcasting and Cable Retransmission, 1993 O.J. (L 248) 15 (EC); Council Directive 93/98 harmonizing the Term of Protection of Copyright and Certain Related Rights, 1993 O.J. (L 290) 9 (EC); Council Directive 96/9 on the Legal Protection of Databases, 1996 O.J. (L 77) 20 (EC); Directive 2001/29/EC of the European Parliament and of the Council of 22 May 2001 on the Harmonisation of Certain Aspects of Copyright and Related Rights in the Information Society (hereinafter EU InfoSoc Directive); Council Directive 2001/84 on the Resale Right for the Benefit of the Author of an Original Work of Art, 2001 O.J. (L 272), 32 (EU).

144 Natali Helberger & P. Bernt Hugenholtz, *No Place Like Home for Making a Copy: Private Copying in European Copyright Law and Consumer Law*, 22 BERKELEY TECH. L.J. 1061,1066 (2007).

145 A few exceptions include the first-sale doctrine, which allows the owner of a copy to resell or otherwise dispose of it, see 17 U.S.C. § 109(a) (2000); the rights of owners of copies of software to make backup copies or adapt them where adaptation "is an essential step in" using the program, see *id.* § 117(a). Similarly, a building owner has

Copyright Act is not to reward authors for their own sake, but through rewarding them to benefit users and society more generally.[146] Despite this recognition of a general user interest, however, very little has been done about the precise shape and scope of this interest. There is not even a consistently used generic term for it. For instance, the US Copyright Act mentions "authors" and "copyright owners" many times, but contains no corresponding generic term to denote those who receive and consume copyrighted works; rather the Act refers to these people variously as "persons",[147] "the public",[148] "owner" of a copy,[149] the "transmission recipient",[150] the "subscriber",[151] and "consumer".[152]

Traditionally, there has been a statutory "free zone" of private use under Chinese copyright law. Within this sphere, users can use the works freely without any intervention by copyright holders. This "free zone" is safeguarded by two important systems – the first sale doctrine and fair use. In this section, I intend to review these two systems. I will also illustrate the restrictions on fair use by a "three-step test" in order to draw a more concrete picture of the legal treatment of private use under Chinese copyright law.

1.2.1 First Sale Doctrine

The first sale doctrine is also called the exhaustion of rights principle. It is "a judicially created limitation on intellectual property rights that judiciously

---

the right under section 120(b) of the Copyright Act to make "alterations to such building, and destroy or authorize the destruction of such building" notwithstanding the exclusive right of adaptation of the copyright owner in the architectural work. *Id.* § 120(b); and some fair use exemptions developed by courts, such as the Sony time-shifting exemption, see *Sony Corp. of Am. v. Universal City Studios, Inc.*, 464 U.S. 417, 429, 454-55 (1984). However, the Supreme Court in Sony failed to recognize a general right of private copying. See Litman, *supra* note 3, at 957, 960-61. A privilege codified in the Audio Home Recording Act authorizes consumers to record music for personal noncommercial use. See Audio Home Recording Act, 17 U.S.C. §§1001-1010 (2000).

146 Liu, *supra* note 48, at 398.

147 See, e.g., 17 U.S.C. § 101 (definition of "publication").

148 See, e.g., *id.* (definition of "publicly").

149 See, e.g., *id.* §§ 109, 117.

150 See, e.g., *id.* § 114(d)(2)(C)(v).

151 See, e.g., *id.* § 119(d)(8).

152 See, e.g., *id.* § 1008.

allows consumers to navigate between the rights of intellectual property owners and the realities of the marketplace."[153]

Although copyright holders have exclusive rights to control the initial dissemination of their works, once published, they lose control over subsequent flow. In other words, the first sale doctrine "assures the copyright owner that, until she parts with ownership, she has the right to prohibit all others from distributing the work. On the other hand, once a sale has occurred, the first sale doctrine allows the new owner to treat the object as his own."[154]

Copyright laws in many countries contain this principle. For instance, Section 109 of the US Copyright Act provides it explicitly:

> Notwithstanding the provisions of section 106 (3), the owner of a particular copy or phonorecord lawfully made under this title, or any person authorized by such owner, is entitled, without the authority of the copyright owner, to sell or otherwise dispose of the possession of that copy or phonorecord.[155]

Similarly, the UK Copyright, Designs and Patents Act 1988 (hereinafter CDPA 1988) provides that any subsequent distribution, sale, hiring or loan of copies previously put into circulation does not infringe the right holders' right to issue or distribute copies of the work to the public.[156]

In comparison, there is no explicit regulation of the first sale doctrine or exhaustion of rights principle in the Chinese IPRs regime. The idea of the first sale doctrine is embodied in the IPRs infringement regulations. For example, Article 69 (1) of the 2009 Chinese Patent Law provides that after the sale of a patented product or a product that was directly obtained by using a patented process by the patentee or by someone with the authorization of the patentee, any other person who uses, offers to sell, sells or imports that product shall not be deemed to have infringed the patent. Although Chinese copyright law does

---

153 Shubha Ghosh, *Recognizing Authority in the Marketplace: The Curious and Ubiquitous Problems of Gray Markets,* in INTELLECTUAL PROPERTY AND INFORMATION WEALTH: ISSUES AND PRACTICES IN THE DIGITAL AGE, VOL.4, INTERNATIONAL INTELLECTUAL PROPERTY LAW AND POLICY 229 (Peter K. Yu ed., Westport: Praeger Publishers 2007).

154 ARTHUR R. MILLER & MICHAEL H. DAVIS, INTELLECTUAL PROPERTY: PATENTS, TRADEMARK, AND COPYRIGHT IN A NUTSHELL 328 (3d ed., US: Thomson/West 2000).

155 17 U.S.C. §109 (a) (2000). For a comprehensive analysis of this Article, see MELVILLE B. NIMMER & DAVID NIMMER, NIMMER ON COPYRIGHT, §8.12[B] (NY: Matthew Bender 1997) (hereinafter NIMMER ON COPYRIGHT).

156 See section 18 (3) (a) of the CDPA 1988.

not contain a similar regulation, the first sale doctrine is generally recognized in the copyright field.[157]

Because of the first sale doctrine, libraries are able to legally lend their collection of copyrighted works to the public,[158] and individuals are able to sell their used books or records on the secondhand market without fear of copyright infringement. However, since the 1980s, copyrights has been expanded to include rental right as a result of the development of reproduction and recording technology, and the applicable scope of the first sale doctrine has shrunk. Fortunately, given that the rental right only refers to specific categories of works such as records, movies, and software, the status of the first sale doctrine has not been fundamentally shifted.[159]

1.2.2 Private Use Exemption within Fair Use

Notwithstanding the fact that the scope and number of copyright exemptions recognized under national positive laws vary widely from country to country, the most common exemptions include quotations, illustrations for teaching, private study and research, criticism or review, reporting of current events, public administration, and justice.[160] Among those exemptions, the one which is most related to the users' benefit from using copyrighted works is the private use exemption.

Article 22 of the 2010 Chinese Copyright Law provides 12 circumstances of fair use exemptions, among which is one instance referring to a private use exemption:

> [A] work may be used without permission from, and without payment of remuneration to, the copyright owner, provided that the name of the author and the title of the work shall be indicated and the other rights enjoyed by the copyright

157 See Han Ning & Li Yuling, Cong zhuzuoquan quanli qiongjie kan zhishi chanquan de xianzhi [*Comment on the Limitations of IPRs from the Exhaustion of Copyright Perspective*], 10 (Mid.) Fazhi yu shehui [LEGAL SYSTEM AND SOCIETY] 348 (2008).

158 See MARY MINOW & TOMAS A. LIPINSKI, THE LIBRARY'S LEGAL ANSWER BOOK 54-57 (US: American Library Association 2003).

159 See Wang Qian, Lun wangluo huanjing zhong de "shouci xiaoshou yuanze" [*"Principles of First Sale" in the Internet Environment*], 3 Faxue zazhi [LAW SCIENCE MAGAZINE] 117, 117 (2006).

160 See e.g. Articles 10 and 10 *bis* of the Berne Convention; Article 5 of the EU InfoSoc Directive; § 29 (1) (research for a non-commercial purpose and private study), § 30 (1) (criticism or review), and § 30 (2) and (3) (reporting current events) of the UK CDPA 1988; 17 U.S.C. § 108-121.

> owner by virtue of this Law shall not be prejudiced: (1) use of a published work of others for the user's own private study, research or self entertainment; ...

Accordingly, the protected private uses should accord with the listed purposes. Any use, as long as it is for the user's own private study, research or self-entertainment purposes, would be exempted from copyright infringement. The requirement of purposes could actually contain almost all manners of using copyrighted works. However, beside the listed ambiguous purposes, there is no further explicit limitation on the amount of use, or any emphasis on the non-commercial nature of the use.

Notably, the existing provision of the private use exemption word-for-word follows the original one contained in the 1990 Chinese Copyright Law.[161] During the period of the 1990s when China's first copyright law was being constituted, private use by the individual user was generally understood to be non-infringing because it is *de minimis*. Due to the lack of personal copying devices at the time and the reliance on handwritten copying by individuals, there was very little effect on the copyright holder's market interest. Private use was therefore recognized as a kind of legitimate fair use.

Even if it was a reasonable choice to exempt all private use from the requirement to obtain permission from or make payment to the copyright holder at that time, however, nowadays the same regulation of a full private use exemption might be problematic in light of the development of mass reproduction technology.[162] To take an example, the reproduction fee in Beijing colleges is very cheap at present. If the market price of a 316-page book is RMB 36, the cost of reproduction is only RMB 14, which is merely 38.88% of the market price.[163]

Undoubtedly, the copy of a whole book would greatly damage copyright holder's legitimate interests. On account of the lack of an explicit restriction on the number of copies that may be made under current copyright law, it might be difficult to prohibit this kind of behavior. Some scholars argue that in order to prevent potential undue harm to the right holder's benefit, it is necessary to

---

161 *Cf.* Articles 22 (1) and (2) of the 1990 Chinese Copyright Law.

162 See Long Wenmao, Zhishi chanquan fazhexue chulun [SURVEY ON THE LEGAL PHILOSOPHY OF INTELLECTUAL PROPERTY LAW] 182 (Beijing: Renmin chubanshe [People's Publishing House] 2003).

163 Peng Biao, Qifen qian tiaozhan zhuzuoquan heli shiyong jiexian [*A Challenge to the Fair Use of Copyright by Seven Cents*], available at http//:zgrdxw.peopledaily.com.cn/gb/paper16/64/class001600001/hwz234465.htm (last visited Jun. 10, 2010).

restrict the manner of fair use, and also the amount of private copying allowed for private study, research or self-entertainment purposes.[164]

1.2.3 Restrictions on Private Use Exemption by "Three-step Test"

Establishing a modern copyright legal framework has been an important part of China's overall policy of opening-up. When drafting the 1990 Chinese Copyright Law and its implementing regulations, legislators were trying to harmonize the law with the relevant requirements of international treaties in order to achieve the normalization of international copyright relationships. Consequently, the general principles and main provisions of the 1990 Chinese Copyright Law and its implementing regulations are basically consistent with the Berne Convention.[165]

For instance, in light of the copyright limitation requirements of the Berne Convention, the 1990 Chinese Copyright Law Implementing Regulations further restricted the fair use of quotation as follows:

> Article 27 The following conditions have to be satisfied for an act to be deemed appropriate quotation of published works by act to be deemed appropriate quotation of published works by others mentioned in Article 22 (2) of the Law: (1) The quotation is made solely for the purpose of introducing or reviewing the source works or making clear a point; (2) What has been quoted does not form a major or substantial part of the work of the quoter; (3) The interests of the copyright owner of the work being quoted shall not be prejudiced.

Regrettably, these restrictions on fair use only refer to quotation. After years of implementation, it has been gradually recognized that other types of fair uses also require certain criteria or restrictions on the manner and amount of use. Thus, the 2002 Chinese Copyright Law Implementing Regulations added a restriction on all 12 fair use circumstances pursuant to the Berne Convention's "three-step test":

> Article 21 The use of a published work without permission from the copyright owner, pursuant to the relevant provisions of the Copyright Law, must not conflict with the normal exploitation of the work and must not unreasonably prejudice the legitimate rights and interests of the copyright owner.

Admittedly, this complementary clause represents a great step forward from what existed previously. However, critics still argue that it only imitates the

164 See Yu Yu & Ji Xiaoxin, *supra* note 141, at 92.

165 See Wei Zhi, *supra* note 60, at 91.

configuration rather than the essence of the "three-step test".[166] The Berne Convention enumerates a series of fair use circumstances in Article 10 (quotations; illustrations for teaching; indication of source and author) and Article 10*bis* (of certain articles and broadcast works; of works seen or heard in connection with current events), and states general exceptions to copyright in Article 9(2)[167] in order to cover the other omitted circumstances. Nevertheless, Article 21 of the 2002 Chinese Copyright Law Implementing Regulations does not function as a general exception to copyright, rather it serves as a further restriction on the provisions of copyright law. In this sense, it neither expands the scope of the 12 enumerated fair use circumstances nor solves the potential danger of fair use enumerations becoming outdated.

In light of Article 21 of the 2002 Chinese Copyright Law Implementing Regulations, the private use exemptions enumerated in Article 22 of the 2001 Chinese Copyright Law must not conflict with the normal exploitation of the work, and must not unreasonably prejudice the legitimate rights and interests of the copyright owner. This is regarded as the fairness standard of private use as well as other exemptions. However, on account of the fact that the standard itself is too ambiguous and difficult to determine, some Chinese scholars recommend a reference to the "four factors"[168] approach under the US copyright law as a supplementary reference.[169]

166 See Lu Haijun, *supra* note 141, at 46.

167 The Berne Convention regulation of exceptions to copyright is usually known as the "three-step test": "It shall be a matter for legislation in the countries of the Union to permit the reproduction of such works in certain special cases, provided that such reproduction does not conflict with a normal exploitation of the work and does not unreasonably prejudice the legitimate interests of the author." See Article 9(2) of the Berne Convention.

168 The four factors of the fair use standard include: (1) the purpose and character of the use, including whether such use is of a commercial nature or is for nonprofit educational purposes; (2) the nature of the copyrighted work; (3) the amount and substantiality of the portion used in relation to the copyrighted work as a whole; and (4) the effect of the use upon the potential market for or value of the copyrighted work. See 17 U.S.C. §107.

169 See e.g., Wu Handong , *supra* note 134, at 194-216; Lu Haijun, Lun heli shiyong zhidu de lifa moshi [*The Legislative Model of the Fair Use System*], 3 Fa shang yanjiu [LAW & COMMERCIAL INDUSTRY RESEARCH] 24 (2007).

1.3 The Impact of Digitisation on the Legal Treatment of Private Use in Contemporary China

The emergence of the Internet has been recognized as a significant technological revolution which has affected society as a whole, and transferred the communication and transmission of information from print media into digital networks. It may have been true in the earlier stages of the Internet that people had the freedom to choose whether or not they were affected by it as suggested by David Pogue when he wrote that, "It doesn't affect people who aren't online (and only a tiny minority of the world population is). And if you don't like the Internet's system, you can always flip off the modem."[170] In the years since, however, the Internet has prevailed and fundamentally affected all aspects of political, economic, social and cultural life. There is no switch that will insulate us from the Internet's effect.

"Copyright law has been seriously challenged by the rise of digital technology, and in particular by the advent and explosive growth of global digital networks."[171] Digitalization of information, on the one hand, facilitates the reproduction and transmission of copyrighted works dramatically, and the above processes have become very convenient and prompt. Digital technology obtains distinct advantages, including its low cost and the extremely limited loss of quality during reproduction, and the whole process is very easily handled by common individuals.

On the other hand, digital technology has changed the way in which intellectual products are distributed and consumed, and makes it possible for instant exchange and simultaneous enjoyment by the masses through P2P networks. To take one example, with no costs of manufacturing, packaging, shipping, or retailing CDs, MP3s online can be offered for download online at a marginal cost of virtually zero and enjoyed by music fans at the cost of just a few minutes of computer time.[172] The new mode of information distribution can be characterized as decentralized and non-mediated. As Antonina Engelbrekt

---

170 David Pogue, *Don't Just Chat, Do Something*, NEW YORK TIMES, Jan. 30, 2000.

171 Pamela Samuelson et al., *The Copyright Principles Project: Directions for Reform*, 25 BERKELEY TECH. L. J. 1, 18 (2010).

172 See Terrel Gallaway & Douglas Kinnear, *Unchained Melody: A Price Discrimination-Based Policy Proposal for Addressing the MP3 Revolution*, 35 JOURNAL OF ECONOMIC ISSUES, 279, 279 (2001).

comments, "[t]he exchange is not B2B and not B2C, but rather C2C, where C stands for both consumer and creator."[173]

In this section, I intend to explore the impact of digitization on the legal treatment of private use in the Chinese copyright regime. As copyright expands dramatically in the digital environment, the statutory "free zone" of private use has not kept pace, but rather has fallen behind. At present, the first sale doctrine is inapplicable in cyberspace, and there is neither a private use exemption within the fair use of information network transmission right, nor a private use exception to the anticircumvention provision of TPMs.

1.3.1 Copyright Expansion in the Digital Environment

As the Internet allows its users to access and share copyrighted files without paying those who have produced them, more and more individual users are becoming the target of copyright infringement litigation. In the pre-digital era, corporate users such as restaurants, hotels and supermarkets, as well as service providers were the traditional favored targets chosen by copyright holders seeking compensation. Individual consumers were conventionally neglected on account of the high transaction costs involved in pursuing claims against them. These dispersed unpermitted behaviors were regarded as too expensive to prohibit.[174]

Originally, litigation against their own consumers did not accord with the interests of copyright holders. When the rampant spread of unauthorized transmission and file sharing by end users through the Internet began to threaten the interests of copyright industries, however, they started to launch lawsuits against individuals as a weapon of last resort.[175]

Just like every coin has two sides, the development of digital technology also presents its pros and cons. As Jessica Litman has observed, on the one hand, the Internet has been hailed as the most revolutionary social development since the printing press. Digital media and network connections are the most democratic form of media, promoting free expression and access to information.

---

173 Antonina Engelbrekt, *Copyright from an Institutional Perspective: Actors, Interests, Stakes and the Logic of Participation*, 4 REVIEW OF ECONOMIC RESEARCH ON COPYRIGHT ISSUES 65, 76(2007).

174 For more details about the market failure consideration of private copying, see *infra* section 3.3.1.

175 See Litman, *supra* note 4, at 1876.

On the other hand, however, "[d]igital technology makes it possible to monitor, record, and restrict what people look at, listen to, read and hear."[176]

Digital technology now permits copyright holders of works in digital format to monitor and meter every single use of their works, and call for a pay-per-view consumption landscape of copyrighted works.[177] Although the corporeal universe has permitted individuals to copy and utilize copyrighted works freely for the purposes of private use, "the on-line era raises the possibility that the publishing industry can track every minuscule use of a work and thereby turn the free use zone into a new opportunity for profit."[178]

Moreover, some scholars advocate authorizing a new "access right" to copyright holders since the exploitation of works has shifted from having copies to directly experiencing the content of the work.[179] As Jane Ginsburg observes, at present the distribution of copyrighted works has moved to an access-based world and thus if the copyright system neglects access control it makes copyright illusory, and in the long run this would be a disservice to consumers.[180]

In spite of concerns for such negative effects, copyright laws in various countries generally refuse to grant copyright holders an "access right". As a substitute, however, they have expanded the scope of copyrights and introduced TPMs to prevent the unauthorized use of copyrighted materials and especially to control the precondition of use – access to the works. Pursuant to the anticircumvention provisions of TPMs, private user's ability to use the copyrighted works is restricted, even for uses that would fall within the statutory fair use exemption. In the subsequent section, I will explore the legal treatment of private use in the digital environment under the Chinese copyright law to illustrate the contraction of permitted private use.

1.3.2 The First Sale Doctrine Is Inapplicable in Cyberspace

With the development and prevalence of digital technology, users have become able to obtain digital copies of works by downloading them from websites. A question that follows from this is whether a user can resell a legally obtained digital version of a work to others under the first sale doctrine. The answer to

---

176 JESSICA LITMAN, DIGITAL COPYRIGHT 12 (NY: Prometheus Books 2001).

177 See *id.* at 13.

178 Marci A. Hamilton, *The TRIPS Agreement: Imperialistic, Outdated, and Overprotective*, 29 VAND. J. TRANSNAT'L L. 613, 615 (1996).

179 See e.g., Ginsburg, *supra* note 43.

180 See *id.* at 124.

this question requires consideration of two issues including – whether the first sale doctrine is applicable in cyberspace, and whether a user's subsequent disposal of works, or a copyright holder's restriction on such a disposal, is legal or not.

Arguably, as a limitation on the distribution right, the reason for the existence of the first sale doctrine is to distinguish between the distribution right of copyright holders and the ownership of the material medium of copyrighted works. Given the fact that the first sale doctrine applies only to the physical or tangible copies of the copyrighted works, rather than the intangible copyright itself, the first sale doctrine is generally held to be inapplicable in the digital environment.

A common approach is to employ another right rather than the distribution right to control the distribution and exploitation of works on the internet. For example, the WIPO Copyright Treaty (WCT) and the WIPO Performance and Phonograms Treaty (WPPT) choose the "right of making available to the public" for the control of digital works, and limit the "right of distribution" only referring to the fixed copies that can be put into circulation as tangible objects.[181] Consequently, the first sale doctrine as a limitation on the distribution right certainly could not apply to the "right of making available to the public" in the digital environment.

Similarly, the EU InforSoc Directive also indicates that the exclusive right to control the distribution of a work is only applicable to a tangible article. Unlike in the analogue world, where the intellectual property is incorporated in a material medium, the disposal of digital copies should be subject to right holders' authorization. Thus the question of exhaustion or the first sale doctrine does not arise in cyberspace.[182]

In contrast, the US copyright law expands the distribution right into the digital environment, and has thus raised fierce domestic debates regarding the application of the first sale doctrine in cyberspace. The libraries camp argues that as long as the user deletes the original copy after transmitting it to others, the first sale doctrine should be applied.[183] This argument has obtained the

181 See Article 6 and Agreed statements concerning Articles 6 and 7 of the WCT; Article 8 and Agreed statement concerning Articles 2(e), 8, 9, 12, and 13 of the WPPT.

182 See EU InfoSoc Directive, Recital Paras. 28, 29.

183 Reply Comments of the Library Associations Before the Library of Congress, The United States Copyright Office and The Department of Commerce, National Telecommunications and Information Administration, Washington, D.C., Sep. 5, 2000, available at http://www.mlanet.org/government/dmca/dmca_reply02.html (last visited Jun. 8, 2009).

support of some American scholars.[184] However, the US Copyright Office has refused to extend section 109 into networks and create a digital first sale doctrine:

> Digital transmission of a work does not implicate the alienability of a physical artifact. When a work is transmitted, the sender is not exercising common-law dominion over an item of personal property; he is exercising the central copyright of reproduction with respect to the intangible work....The underlying purposes of the first sale doctrine is to ensure the free circulation of tangible copies.[185]

In China, when the copyright law was revised in 2001, a new "information network transmission right" was created in order to control the transmission and exploitation of works on the internet. This approach means that the first sale doctrine is inapplicable in cyberspace too.[186]

1.3.3 No Private Use Exemption within Fair Use of Information Network Transmission Right

On the eve of entering the WTO, China revised its copyright law and added a new exclusive right – the information network transmission right – for copyright holders.[187] At the same time, the State Council was empowered to constitute regulations concerning specific protections for this new copyright.[188] Arguably, the revision, to some extent, was influenced by the WCT and the TRIPS. For instance, the right of information network dissemination was created in order to integrate the regulations in international treaties within the domestic legal culture, and to attach importance to copyright protection in the new digital environment.

This innovative action to create a new category of copyright has been praised by many Chinese scholars. They consider it to be much clearer and more concrete than including it in any other rights, such as the "right of communication to the public", the "right of making available", or a broader

---

184 E.g., R. Anthony Reese, *The First Sale Doctrine in the Era of Digital Networks*, 44 B. C. L. REV. 577 (2003).

185 US Copyright Office, DMCA: Section 104 Report (A Report of the Register of Copyrights Pursuant to Section 104 of the Digital Millennium Copyright Act) (2001) 87, available at http://www.copyright.gov/reports/studies/dmca/dmca_study.html (last visited Jun. 15, 2010).

186 See Wang Qian, *supra* note 159, at 119.

187 See Article 10 (12) of the 2001 Chinese Copyright Law.

188 See *id.* Article 58.

"right of reproduction".[189] However, there has been a fierce debate regarding whether to create such a new right. Some Chinese government officials, judges, and scholars considered that the previous "existing regulations could be applied directly to the copyright infringements on the Internet", or "at least could be regulated by the 'etc' circumstances in the current law".[190] These arguments were not, however, accepted by legislators in the end.

Arguably, it is necessary to establish a new right of information network transmission in China. The current reproduction right and distribution right, or the combination of both rights together, could not cover the connotation of the information network transmission right. Conventionally, China has adopted a narrow scope of the reproduction right, which only referred to a process of reproducing information consistent with the original, and failed to protect various representations of "heterogenous reproduction".[191] In respect of the distribution right, it is the right to provide the public with original copies or reproduced copies of works.[192] It is different from network transmission in the channel or manner of transmission. Thus, the reproduction right or distribution right, or both, could not fully contain the character of information network transmission in respect of providing "the public with works by wired or wireless means", and making "the public able to respectively obtain the works at the individually selected time and place".[193]

Similar to other exclusive copyrights, the information network transmission right is also subject to copyright limitations. Article 6 of the Regulations for the Protection of Information Network Transmission Right (hereinafter the 2006 Transmission Regulation) enumerates eight fair use exceptions to the information network transmission right. In contrast with Article 22 of the 2001

189 See e.g., Qiao Sheng, Xinxi wangluo chuanbo quan lifa pingjia yu wanshan [*Evaluation of the Lawmaking Process of the Information Network Transmission Right and Its Improvement*], 4 Zhongguo faxue [CHINESE LEGAL SCIENCE] 15, 15 (2004).

190 See Zhang Yurui, Hulianwang shang zhishi chanquan – susong yu falü [INTELLECTUAL PROPERTY ON THE INTERNET: LITIGATION AND LAW] 19 (Beijing: Renmin fayuan chubanshe [Court Press] 2000).

191 See Liu Chuntian ed., Zhishi chanquan fa [INTELLECTUAL PROPERTY LAW] 58 (Beijing: Zhongguo renmin daxue chubanshe [China Renmin University Press] 2000). *Cf.* Article 10(5) of the 2001 Chinese Copyright Law ("the right of reproduction, that is, the right to produce one or more copies of the work by means of printing, Xeroxing, rubbing, sound recording, video recording, duplicating, or re-shooting, etc.").

192 See Article 10(6) of the 2010 Chinese Copyright Law ("the right of distribution, that is, the right to provide the public with original copies or reproduced copies of works by means of selling or donating.").

193 See *id.* Article 10 (12).

Chinese Copyright Law, which provides twelve circumstances of fair use, several exemptions have disappeared.[194]

Concerning private use, Article 22 of the 2001 Chinese Copyright Law includes a private use exemption referring to the use of a published work for private study, research or self entertainment purposes. However, it is excluded from the eight fair use exemptions to the information network transmission right in Article 6 of the 2006 Transmission Regulation. As a result, the conventional statutory "free zone" of private use is abandoned under the digital copyright regime.

1.3.4 No Private Use Exception to the Anticircumvention Provision of TPMs

In order to protect copyright holders' interests in the digital environment and prevent unauthorized reproduction, sharing, and transmission of copyrighted works, TPMs are recognized by copyright laws in most countries. Because technological relief could prevent copyright infringements *ex ante*, it might be more efficient and effective against piracy than traditional legal relief *ex poste*, especially in the digital environment where infringements have become easier and larger in scale.[195]

The WCT and the WPPT contain obligations concerning technological measures and rights management information to protect copyright in the digital era.[196] Accordingly, despite its being noncompulsory for member states to impose the use of TPMs, which is left at the discretion of right holders, the contracting parties are required to provide adequate legal protection and effective legal remedies against the circumvention of "effective" technological measures that are used by copyright holders.

Before adopting TPMs provision in statute, the notion of protecting TPMs had already existed in China. For example, the first litigated infringement case of TPMs circumvention in China was *China Academy of Building Research v. Zhang Lisun & Chongqing Computer Week* in 1998.[197] Although there were no

194 E.g., *id.* Articles 10 (1), (9) and (10).

195 See Wang Qian, Dui jishu cuoshi lifa baohu de bijiao yanjiu [*Comparative Study on the Protection of Technological Protection Measures*], 2 Zhishi chanquan [INTELLECTUAL PROPERTY] 3, 3 (2003).

196 See Articles 11 and 12 of the WCT, and Articles18 and 19 of the WPPT.

197 Defendant Zhang Lisun developed software, the sole purpose of which was to decrypt the copyrighted software of the plaintiff. The defendant displayed advertising of the decryption software on the *Chongqing Computer Week*. See Yan Xinhua, Ruanjian feifa

legislative regulations to prohibit the circumvention of TPMs when the case appeared, the Chongqing No.1 Intermediate People's Court made a verdict that the defendant's circumvention behavior of TPMs violated copyright law.

As the prevalence of TPMs adoption by copyright holders in digital network transmissions became greater the need to protect, TPMs in statute became more urgent. Although at the time China had not entered the WCT and the WPPT, there was a trend towards harmonizing with these international treaties sooner or later. Thus, when revising the copyright law in 2001, legislators introduced the anticircumvention provision of TPMs from international treaties. Article 47 (6) of the 2001 Chinese Copyright Law provides that without permission from the copyright owner, intentionally avoiding or destroying the technical measures taken by the right holder on works to protect the copyright or the rights related to the copyright will result in civil, administrative, or even criminal liability.

In order to implement the basic principle of protecting the information network transmission right in copyright law, China enacted the special 2006 Transmission Regulation and provided anticircumvention of TPMs in detail.[198]

Notably, the 2006 Transmission Regulation does not distinguish between the protection standards applied to different technological measures.[199] Article 4 states that it not only prohibits the making or selling of devices or services that are used to circumvent, but also prohibits the act of circumvention itself. Articles 18 and 19 further provide the civil, administrative and criminal liability for circumventing TPMs.

Consequently, it is generally forbidden to circumvent the TPMs adopted by copyright holders. Any circumvention except where otherwise provided for in laws or administrative regulations would constitute copyright infringement. Article 12 of the 2006 Transmission Regulation provides four exceptions to the circumvention prohibition, including the nonprofit educational institution or scientific research exception, nonprofit uses for the benefit of the blind, administrative or judicial activities, and security testing. However, the 2006 Transmission Regulation fails to clarify the relationship between TPMs and fair use. Besides these four narrow enumerated exceptions to the anticircumvention provision, there is no acknowledgement of a fair use exemption in general.[200]

---

jiemi qinquan [*Illegal Software Decryption Constitutes Copyright Infringement*], Keji ribao [SCIENCE AND TECHNOLOGY DAILY], Nov. 15, 2000.

198 See Article 4 of the 2006 Transmission Regulation.

199 By contrast, the DMCA distinguishes the access control measures and copyright protection measures, and accordingly applies different protection. See 17 U.S.C. § 1201.

200 See Feng Xiaoqing, Jishu cuoshi yu zhuzuoquan baohu tantao [*Research on Technological Protection Measures and Copyright Protection*], 4 Faxue zazhi [LAW SCIENCE MAGAZINE] 20, 37 (2007) (stating that the four fair use exemptions of

Among these four exceptions to the anticircumvention rules, none of them cover any private use. In this sense, as long as the copyright holders employ TPMs in their copyrighted works, private users are not allowed to circumvent the blockage even for the legitimate fair use acknowledged in the positive law. As a result, together with the lack of recognition of a private use exemption to the information network transmission right, there is no "free zone" of private use under the double sieges of the law and the technological measures.

1.4 Summary

In this chapter, I have discussed the basic framework of Chinese copyright law, the legal treatment of private use under Chinese copyright law, and especially the impact of digitization.

The Chinese copyright law protects the right holder's moral rights and economic rights. Any one who commits infringement shall bear civil, administrative, or criminal liability. In respect of copyright implementation, administrative agencies together with the courts form the so-called copyright "dual system".

Notably, copyright protection is not the sole objective of Chinese copyright law. It also serves for the society's public interests in the free flow of information, and dissemination of works which would contribute to the development and flourishing of socialist culture and sciences. The copyright limitation system, especially the fair use, is an important built-in balancing tool for this purpose.

China has adopted an exhaustive list of fair uses enumerating 12 circumstances, among which one instance refers to a private use exemption. Any use as long as it is for the user's own private study, research or self-entertainment purposes, and further passes a "three-step test" would be exempted from copyright infringement. Fair use together with the first sale doctrine shields a statutory "free zone" of private use.

As copyright expands dramatically in the digital environment, however, the statutory "free zone" of private use has not kept pace, but has rather fallen behind. At present, the first sale doctrine is inapplicable in cyberspace, and there is neither a private use exemption within fair use to the information network transmission right, nor a private use exception to the anticircumvention provision of TPMs. The overprotection of TPMs without considering users' legitimate interests disrupts the conventional delicate balance between right holders' exploitation and users' enjoyment.

---

anticircumvention provision of TPMs in the 2006 Transmission Regulation is too narrow and needs to be improved).

Chapter II Private Use and the China Piracy Battle

In order to solve a certain problem, we need to expose its real causes or ultimate factors. Arguably, the anticipated purpose of not having a private use limitation in the Chinese TPMs framework is as an anti-piracy measure. However, it is in fact a "wrong remedy" against piracy.

On the one hand, it fails to catch the deep-seated causes of the piracy dilemma in China. There are a range of factors known as "blocks" hindering the settlement of the long term piracy problem in China. These are to be found in the law making process, the implementation of the law, the level of adherence to the law, and also in the market.

On the other hand, private use and piracy are different in essence.[201] Private copies are used for non-commercial private purposes and limited to the private sphere, while pirated copies enter into public distribution and are normally for commercial exploitation.

In this section, I first explore the legislative background behind the exclusion of a private use limitation from the digital copyright framework, and further investigate the real causes of the piracy dilemma in China.

2.1 Legislative Background of the Chinese Digital Copyright System

Compared with what the law is, perhaps a more important investigation should focus on why the law regulates in the way it does. This section examines the legislative background of the 2006 Transmission Regulation which forms the main body of the Chinese digital copyright regime.

2.1.1 Great Pressure Against China's Piracy

As Hillary Clinton declared during the 2008 presidential election, "I do not want to eat bad food from China … or have my children having toys that are going to get them sick."[202] The round of China-bashing not only referred to the safety of

201 See *infra* section 3.1.

202 The TIMES, *Democrats Found Wanting as China-bashing Comes to Fore*, Aug. 9, 2007, available at http://www.timesonline.co.uk/tol/comment/columnists/bronwen_maddox/article2224879.ece (last visited Aug. 16, 2010).

numerous products made in China, but also quite often included the long-term copyright piracy puzzle.

A series of headlines from one mainstream media outlet– Business Week magazine – provided evidence of the screeching China-bashing, including headlines such as *Piracy Problem Sails to Rural China;*[203] *Deaf to Music Piracy: Chinese Search Engines Make It Easy to Steal Net Tunes;*[204] *How to Win the China Piracy Battle;*[205] *U.S. Takes Piracy Pushback to WTO – Intellectual-Property Rights Violations in China Cost the U.S. Billions Each Year, Leading to Complaints to the World Trade Organization;*[206] *Chinese Fakes: Tough to Police;*[207] *China: Putting a Stop to IP Piracy.*[208]

Just as earlier in the 1930s, it was observed that "[t]here is probably more misinformation about China than about any other country in the world,"[209] copyright piracy is empirically confirmed to be a world wide, rather than a specifically Chinese, problem.[210] The widespread perception that copyright

---

203 Victoria Ho, *Piracy Problem Sails to Rural China*, BUSINESS WEEK, May 15, 2008, available at http://www.businessweek.com/globalbiz/content/may2008/gb20080515_262289.htm (last visited Aug. 16, 2010).

204 Bruce Einhorn & Xiang Ji, *Deaf to Music Piracy: Chinese Search Engines Make It Easy to Steal Net Tunes*, BUSINESS WEEK, Aug. 30, 2007, available at http://www.businessweek.com/globalbiz/content/aug2007/gb20070830_502991.htm?chan=search (last visited Jun. 10, 2010).

205 Shaun Rein, *How to Win the China Piracy Battle*, BUSINESS WEEK, Jun. 20, 2007, available at http://www.businessweek.com/globalbiz/content/jun2007/gb20070620_006304. htm?chan=search (last visited Jun. 12, 2010).

206 Frederik Balfour, *U.S. Takes Piracy Pushback to WTO – Intellectual-Property Rights Violations in China Cost the U.S. Billions Each Year, Leading to Complaints to the World Trade Organization*, BUSINESS WEEK, Apr. 10, 2007, available at http://www.businessweek.com/globalbiz/content/apr2007/gb20070410_466097.htm?chan=search (last visited Jun. 12, 2010).

207 Frederik Balfour, *Chinese Fakes: Tough to Police*, BUSINESS WEEK, Apr. 9, 2007, available at http://www.businessweek.com/globalbiz/content/apr2007/gb20070409_915993. htm?chan=search (last visited Jun. 13, 2010).

208 Alistair J. Nicholas, *China: Putting a Stop to IP Piracy*, BUSINESS WEEK, Mar. 22, 2007, available at http://www.businessweek.com/globalbiz/content/mar2007/gb20070322_286139.htm?chan=search (last visited Jun. 13, 2010).

209 EUNICE TIETJENS & LOUISE STRONG HAMMOND, BURTON HOLMES TRAVEL STORIES: CHINA 209 (Chicago: Wheeler Publishing Co. 1931).

210 Copyright piracy in developed countries is also a great problem. E.g., Spain is viewed by the IFPI as "Europe's most serious piracy problem country where rampant street CD piracy has shrunk the legitimate market by one third in the last three years". IFPI, *One in Three Music Discs Is Illegal but Fight Back Starts to Show Results*, Jun. 23, 2005, available at http://www.ifpi.org.uk/content/section_news/20050623.html (last visited

piracy is only or primarily a problem in developing countries, and especially East Asian countries like China, has recently been shown to be incorrect and without sufficient supporting evidence.[211] Even if research could show that China is not the No. 1 country for piracy in the world, as it is all to frequently portrayed, the emerging problem of piracy still deserves deeper investigation.

The anti-piracy issue remains an unexceptional topic in the annual Chinese government work reports[212] and intellectual property rights protection White Papers.[213] Every year, the Chinese government devotes huge amounts of agency resources, financial funds as well as human resources against piracy. The anti-piracy task has obtained great achievement. In the year 2008, for example, China investigated and dealt with 12,490 piracy cases and captured 76,055,000 pirate publications, among which 56,795,000 were pirate audiovisual products, 13,094,000 were pirate books, 3,351,000 were pirate teaching materials, and 2,815,000 were pirate software and electronic publications. On April 26, 2008, World Intellectual Property Day, the National Anti-Pornography and Anti-Illegal Publications Office deployed to 31 provinces and cities at the same time to collectively destroy a total of 47,180,000 pirated and other illegal publications.[214]

Jun. 15, 2010); Canada has been accused of being the origin of a half of the world's pirated movie recordings. See Vito Pilieci, *Hollywood Blames Canada for Half of Movie Piracy*, Canada.com, Jan. 24, 2007, available at http://www.canada.com/topics/technology/story.html?id=f8ae08f5-b82d-4e87-97b9-67ff7097638f&k=65095 (last visited Jun. 15, 2010).

211 See Aaron Schwabach, *Intellectual Property Piracy: Perception and Reality in China, the United States, and Elsewhere*, 2 JOURNAL OF INTERNATIONAL MEDIA AND ENTERTAINMENT LAW 65,65 (2008). *Cf.* e.g., Peter K. Yu indicates four common misconceptions about copyright piracy: "(1) copyright piracy is merely a cultural problem, (2)copyright piracy is primarily a development issue, (3) copyright piracy is a past necessary phenomenon for technologically-advanced countries, and (4) copyright piracy is a byproduct of authoritarian rule." See Peter K. Yu, *Four Common Misconceptions about Copyright Piracy*, 26 LOY. L. A. INT'L & COMP. L. REV. 127, 130 (2003).

212 Every year at the National People's Congress, the premier will present the report on the work of the government on behalf of the State Council. Among all government work, intellectual property protection is a very important task.

213 Since 1998, the State Intellectual Property Office (SIPO) issues the annual White Paper on China's Intellectual Property Rights Protection, available at http://www.sipo.gov.cn/sipo_English/laws/whitepapers/ (last visited Feb. 13, 2010).

214 2008 nian "sao huang da fei" gongzuo qude xin de tupo [In 2008 the task of "Anti-Pornography and Anti-Illegal Publications" obtained new breakthrough], available at http://www.gapp.gov.cn/cms/html/47/711/200902/462577.html (last visited Jun. 15, 2010).

However, copyright piracy repeatedly revives [215] after various such "movements".[216] Traditional anti-piracy measures mainly aim at tangible pirates. One mediagenic image is the strike against pirated DVD: card tables covered with the latest Hollywood movie DVDs, greedy street vendors being led away in handcuffs, thousands of accumulated videos being crushed down and set on fire. [217] With the prevalence of Internet, however, the piracy of digital copyrighted works has become increasingly difficult to pursue, which makes the situation of copyright protection even more complicated.

The last ten years has been a golden age for the rapid development of the Chinese computer industry and for the popularization of the Internet. By the end of June, 2008, the number of netizens in China had reached 253 million, and the size of the netizen population in China has leaped to first place in the world.[218] Meanwhile, byproducts of digitalization, such as counterfeit software, bootlegged films online, free-downloaded MP3 and peer-to-peer file sharing are becoming the new hot potatoes on the menu of copyright protection. Therefore, protecting copyright industries' economic interests in the digital environment has become a priority during the amendment of Chinese copyright law.

2.1.2 Impact from the International Copyright Protection Framework

Keeping in line with international IPR conventions and normalizing international copyright trade relationships are important parts of China's opening-up policy.[219] From the establishment of its first copyright law, China has always emphasized the need to harmonize its domestic rules with the international copyright protection framework. When revisions to the 1990

---

215 See Liu Wenjing, WTO guize guonei shishi de xingzheng fa wenti [ADMINISTRATIVE LAW ISSUES IN THE IMPLEMENTATION OF WTO REGULATIONS] 73 (Beijing: Beijing daxue chubanshe [Peking University Press] 2004).

216 For instance, in 2006 the "Blue Exhibition Action" and "Anti-piracy Every Day Action"; in 2008, the "Thunder Storm" and "Skynet" Movements as well as long-term combined enforcements to protect Olympic marks.

217 See, e.g., *Police Destroy 2.5 Million Fake VCDs, DVDs*, JAKARTA POST, Sep. 20, 2007, available at http://www.thejakartapost.com/detailgeneral.asp?fileid=20070912143534&irec=69# (last visited Jun. 15, 2009).

218 See China Internet Network Information Center (CNNIC), *The 22nd Statistical Survey Report on the Internet Development in China*, Abridged Edition (July, 2008), available at http://www.cnnic.cn/download/2008/CNNIC22threport-en.pdf (last visited Jun. 18, 2010).

219 See Wei Zhi, *supra* note 60, at 91.

Chinese Copyright Law were being discussed in order to adapt to new situations in the digital environment, the two main treaties – the WCT and the WPPT – which constitute the international copyright protection framework were used for reference.

However, these international models which China intends to draw lessons from are not always "the avatar of justice". Arguably, the existing international copyright framework and expansive tendency mainly represent the interests of few giants headed by the US.

Before the establishment of the WCT and the WPPT, in the domestic domain, the Clinton White Paper[220] which reflected an expansive copyright protection was rejected by the US Congress in the first instance. Given fierce resistance from consumer groups, it was very difficult to persuade Congress to pass the Clinton White Paper. By contrast, it would be much easier during the process of passing international treaties. Therefore, the US copyright industries changed their strategy and transferred their battlefield into the international arena instead.

During the negotiation process of the WCT and the WPPT, the US delegation advocated a proposal similar to the Clinton White Paper, and ultimately convinced WIPO to adopt it. After the treaties went into effect, it became the US government's obligation to implement them on the domestic level. As the US revised its copyright law in light of the WCT and the WPPT, US copyright interest groups finally obtained the benefit they had failed to gain in Congress. Through dominating the international copyright protection framework, US copyright industries have not only won the battle in the domestic arena against national consumer groups, but have also exported their own protection policy and standards to the rest of the world.[221]

The reason that the US delegation proposed very expansive and strong copyright protection into international treaties was these interests. It is in accordance with the US's national interests to fully protect the American copyright industries' benefits. In the US, the copyright based industries continue to be one of America's largest and fastest-growing economic assets. "The U.S.

---

220 Working Group on Intellectual Property Rights, US Dep't of Commerce, Intellectual Property and the National Information Infrastructure: The Report of the Working Group on Intellectual Property Rights, Sep. 1995 [hereinafter the Clinton White Paper], available at http://www.uspto.gov/web/offices/com/doc/ipnii/ (last visited Jun. 18, 2010).

221 In addition to the use of international treaties, the US also has made great efforts to spread its domestic copyright rules to other countries through free trade agreements, for instance the Australia-United States Free Trade Agreement (AUSFTA) exports the US DMCA regime to Australia.

copyright industries have consistently outperformed the rest of the U.S. economy, in terms of contribution to gross domestic product and employment levels, and have played an increasingly prominent role in the growth of U.S. exports."[222]

In 2002, the US "total" copyright industries accounted for an estimated $1.25 trillion or 12% of the entire US economy, among which, the "core" copyright industries[223] amounted to $626.6 billion or 6% in GDP.[224] In 2007, the core copyright industries contributed $889.1 billion to US GDP, equal to approximately 6.4% of the US economy.[225]

Meanwhile, intellectual property is now the largest export industry in the US.[226] Large numbers of cultural products that embody American creativity, such as motion pictures, television programmings, computer software, books, newspapers and periodicals, are sold throughout the world. According to an estimate by the Motion Picture Association of America (MPAA), foreign sales revenue for all media in 2003 reached $19 billion.[227]

Copyright industries – such as music, film, television, and computer software – are central pillar to the health and stability of the US economy. There was no doubt that the US would adopt a copyright policy strongly partial to copyright holders. By virtue of its powerful influence, the US has successfully extended its model to the whole world.

However, the US copyright protection framework is not always suitable to other countries. Institutional choice is contingent on the historical and

---

222 World Intellectual Property Organization (WIPO), The Economic Contribution of Copyright-Based Industries in USA: The 2004 Report [hereinafter The 2004 Report], Economists Incorporated 2004, p.17, available at http://www.wipo.int/export/sites/www/ip-development/en/creative_industry/pdf/ecostudy-usa.pdf (last visited Sep. 8, 2010).

223 The definition of the "core" industries has been expanded to reflect the WIPO definition. In The 2004 Report, the "core" industries are those industries whose primary purpose is to produce or distribute copyright materials. These industries include newspapers, book publishing, recording, music, and periodicals, motion pictures, radio and television broadcasting, and computer software (including business application and entertainment software). See *id.* at 3.

224 See *id.* at 1.

225 See RIAA, *About Music Copyright Notices*, available at https://www.riaa.com/ispnoticefaq.php (last visited Sep. 12, 2010).

226 See Paul Farhi & Megan Rosenfeld, *American Culture Makes It a Small World, After All*, STAR TRIBUNE (Minneapolis, Minn.), Nov. 26, 1998, at A27 (indicating that according to an estimation that international sales of software and entertainment products were worth $ 60.2 billion in 1996, more than any other United States industry).

227 See The 2004 Report, at 16.

institutional context and is often country-specific.[228] Other jurisdictions should always adopt a sensible response to US effort to export its standards of protection, whether it is directly through free trade agreements, or indirectly through international treaties.

Turning to the backdrop of Chinese copyright law revision, in 2001 China revised the statute to meet the requirements for entering the WTO. Subsequently, China promulgated the 2006 Transmission Regulation in order to implement the obligations of the WCT and the WPPT.

This series of revisions, to a large extent, are institutional transplants from the international copyright protection framework and passive acceptance of established arrangements made by a few foreign giants headed by the US, rather than a deliberate choice based on the realistic Chinese national situation. Moreover, when implementing the obligations of international treaties, China has adopted an advanced copyright protection level which far exceeded the minimum requirements of the WCT and the WPPT. In the absence of a full evaluation of the effect on domestic copyright industries and the market, this action seemed unadvisable.

2.1.3 Piracy Suspected P2P File Sharing by Individual Users

P2P technology has transformed the conventional user's experience of getting information via the Internet. "P2P is a disruptive technology for deploying applications that scale to millions of simultaneous participants."[229] As David Barkai indicates: "[t]he idea behind P2P computing is that each peer, i.e., each participating computer, can act both as a client and as a server in the context of some application."[230] For instance, in the case of P2P networks like Napster, KaZaA, Gnutella, and Morpheus, users maintain a shared collection of files on their computers.[231] Via a P2P network, they may search and download files from

---

228 Engelbrert, *supra* note 173, at 71.

229 JOHN F. BUFORD, HEATHER YU, & ENG KEONG LUA, P2P: NETWORKING AND APPLICATIONS 23 (USA: Morgan Kaufmann Publishers 2009).

230 DAVID BARKAI, PEER-TO-PEER COMPUTING: TECHNOLOGIES FOR SHARING AND COLLABORATING ON THE NET 4 (Hillsboro, OR: Intel Corporation 2002).

231 E.g., a popular P2P software in China –DC++ requires that the user at least retains a 2G shared collection, which is equivalent to the size of 600 MP3 music files. Another prevalent P2P software 100Bao demands an automatic setting up the downloading files as shared ones.

other users' shared collection, and also let others search and download from their own shared collections.

Around 2000, "Napster quickly became the world's largest community for sharing music files because it allowed easy searching, had a user-friendly interface, let the users communicate with each other through instant messaging and chat rooms, and enabled them to share each other's bookmarks." [232] According to Howard Rheingold, "[a]t its height, 70 million users were trading 2.7 billion files per month".[233]

Since their emergence, P2P networks have been highly controversial. As Alexander Peukert points out:

> Peer-to-peer networks provide architecture for stable, cheap, and global sharing of any digitized information, be it music, movies, software, writings or other data. The end-to-end or peer-to-peer architecture makes it possible for thousands of terabytes to rush through P2P networks every month without anybody having to invest in and provide for a centralized server. The technology features characteristics that prompt great hopes for the advent of the global knowledge community. However, it also terrifies copyright owners to definitely lose control over their works, which for the user of these networks actually seem to be "free as the air to common use".[234]

P2P file sharing enables the spread of content for which the right holder has not been paid, and induces a war between copyright holders, and end users and P2P service providers. [235] However, not all P2P file sharing is harmful. [236] As

---

232 HOWARD RHEINGOLD, SMART MOBS: THE NEXT SOCIAL REVOLUTION 72 (Cambridge, MA: Perseus Publishing 2002).

233 *Id.* Paul Goldstein makes a similar observation, see PAUL GOLDSTEIN, COPYRIGHT'S HIGHWAY: FROM GUTENBERG TO THE CELESTIAL JUKEBOX 106 (revised edition, California: Stanford University Press 2003) ("At its height, the Napster music file sharing service claimed to have over 70 million subscribers.").

234 Alain Strowel, *Introduction*, citing Alexander Peukert' argument in *A Bipolar Copyright System for the Digital Network Environment*, in PEER-TO-PEER FILE SHARING AND SECONDARY LIABILITY IN COPYRIGHT LAW 2 (Alain Strowel ed., UK: Edward Elgar Publishing Limited 2009).

235 See Peukert, *id.* at 149 (indicating that P2P networks have already given rise to a war "between copyright owners, and users and their proponents, who want to retain the benefits of this technology for themselves and for society at large.").

236 See Netanel, *supra* note 6 ("P2P file sharing is not just downloading music and movies for free. It is a vehicle for finding works that are otherwise not available, discovering new genres, making personalized compilations, and posting creative remixes, sequels, and modifications of popular works. By engaging in such activities, people who might previously have been passive consumers now assert a more active, self-defining role in the enjoyment, use, and creation of cultural expression. They also share their interests, creativity, and active enjoyment with others.").

Lawrence Lessig sums up, there are four different categories of sharing in the case of music:

> A. There are some who are using sharing networks as substitutes for purchasing CDs.
>
> B. There are also some who are using sharing networks to sample, on the way to purchasing CDs.
>
> C. There are many who are using file-sharing networks to get access to content that is no longer sold but is still under copyright or that would have been too cumbersome to buy off the Net.
>
> D. There are many who are using file-sharing networks to get access to content that is not copyrighted or to get access that the copyright owner plainly endorses.[237]

Accordingly, he suggests that the law should differentiate the treatments towards diverse file sharing:

> The Internet is in transition. We should not be regulating a technology in transition. We should instead be regulating to minimize the harm to interests affected by this technological change, which enabling, and encouraging, the most efficient technology we can create. We can minimize that harm while maximizing the benefit to innovation by
>
> 1. guaranteeing the right to engage in type D sharing;
>
> 2. permitting noncommercial type C sharing without liability, and commercial type C sharing at a low and fixed rate set by statute;
>
> 3. while in this transition, taxing and compensating for type A sharing, to the extent actual harm is demonstrated, possibly limiting the tax to those who choose to download music.[238]

Similarly, several commentators have suggested a solution where non-commercial P2P file sharing would be allowed in exchange for the imposition of a levy or a tax on P2P related services and products.[239] However, in a Chinese context, there is no such discriminating treatment towards different kinds of P2P file sharing.

---

237 LAWRENCE LESSIG, FREE CULTURE: THE NATURE AND FUTURE OF CREATIVITY 296-97 (NY: Penguin books 2004).

238 *Id.* at 303.

239 See e.g., Netanel, *supra* note 6; WILLIAM W.FISHER III, PROMISES TO KEEP: TECHNOLOGY, LAW, AND THE FUTURE OF ENTERTAINMENT (California: Stanford University Press 2004).

2.1.4 No "Free Zone" of Private Use in Digital Copyright Framework as an Anti-piracy Measure

As indicated in the first chapter, there is no "free zone" of private use in the digital copyright framework. Private use is excluded from both fair use of the information network transmission right, and the exceptions to the anticircumvention provisions of TPMs.[240]

Arguably, the cause of such a treatment towards private use is the outcome of anti-piracy activities. Out of the great pressure against piracy from both domestic copyright industries and international societies, the Chinese government not only provides a high level of copyright protection in statue, but has also devoted a huge amount of agency forces, financial funds and human resources against piracy.[241] There is therefore little wonder that where there remains uncertainty on private use the Chinese legislator will lean towards protecting right holders' interests, as will be discussed in the next section.

The vague concept of private use contains a danger of confusion and misunderstanding in China. According to the Chinese copyright law, any use would be exempted from copyright infringement as long as it was for the user's own private study, research or self-entertainment purposes.[242] This broad description of private use could contain almost all ways of using copyrighted works.

On account of the vague concept of private use and the lack of particular eligibility criteria for determining what constitutes a private use, protections of a private use exemption might be abused. For example, in respect of P2P file sharing there are fierce debates about this. Some users who share files via a P2P network have argued that their sharing is for the purposes of private study, research, or self-entertainment, and belongs in the category of fair use.[243] In comparison, right holders have argued that large-scale unauthorized P2P file

---

240 See *supra* sections 1.3.3 & 1.3.4.

241 See *supra* sections 2.1.1 & 2.1.2.

242 See Article 22(1) of the 2001 Chinese Copyright Law.

243 See Wang Qian, P2P ruanjian zuizhong yonghu banquan qinquan wenti yanjiu [*Research on End User Liability with P2P Software*], 5 Zhishi chanquan [INTELLECTUAL PROPERTY] 9, 11 (2004) (indicating that the Chinese copyright statute neither illustrates that the works in case of private use must be legal versions, nor as the US copyright law does, sets out subsequent factors for determining fair use. According to a literal interpretation of Article 22 (1) of the 2001 Chinese Copyright Law, downloading digital works via P2P network for private use purposes does not violate copyright.).

sharing undermines their economic rewards, and constitutes illegal copyright infringement.[244]

Despite the great convenience that P2P file sharing offers users, it is also a fact that right holders' economic rewards have thereby greatly decreased. A large number of end users are now able to share copyrighted works, such as MP3 music, movies, and digital books. With no remuneration being paid to right holders, P2P file sharing undermines the copyright incentive. As Alexander Peukert points out:

> If a single digital copy made available on the Internet, especially in a peer-to-peer network, theoretically suffices as a master for an indefinite number of identical, unauthorized clones, right holders fear that they will not be able to recoup their investment in producing the work in the first place. Without this promise, it is argued, fewer works will be created.[245]

This is especially apparent in the record industry. According to a survey by the International Federation of the Phonographic Industry (IFPI), in April 2003, there were 1 billion instances of file-sharing of unauthorized music. In the three years from 2001 to 2003, music sales in the world fell 14%. Between 1997 and 2003, retail sales of CD and cassette sound recordings decreased 30% in both Canada and Germany. The value of the Canadian market lost more than $ 425 million, and the German market lost 800 million euro. In Denmark, sales of CD albums fell nearly 50% from 2001-2003. Surveys in all major markets have shown that P2P file sharing was the major factor in the fall of record sales worldwide.[246]

Finally, at the moment when Chinese legislators constituted the 2006 Transmission Regulation, there remained a lot of confusion as to the boundary between private use and piracy both in theory and in judicial practice. Given the fact that pirates might make use of a private exemption as a safe harbor, legislators adopted an oversimplified and crude way to exclude private use from the scope of fair use of digital copyright, rather than taking a more careful

---

244 See *A & M Records Inc. v. Napster Inc* 239 F. 3d 1004 (9th 2001); Wang Qian, *id.* at 9 (arguing that a large number of downloading and providing copyrighted works via P2P networks severely harms copyright holders' benefits. Downloading copyrighted works via P2P networks constitutes "copying", and can hardly be exempted as fair use or under a private use exception. Besides, maintaining a shared collection of copyrighted works on their computers for others to search and download constitutes infringement of the information network transmission right. These copyright infringements cannot obtain legality on account of the difficulty to charge end user liability.).

245 Peukert, *supra* note 234, at 148.

246 IFPI, *Fact Sheets-Internet Piracy*, available at http://www.ifpi.org/content/section_news/ 20040330c.html (last visited Sep. 12, 2010).

balancing approach which differentiated among various categories of P2P file sharing.

2.2 The Gap between "Law on the Books" and "Law in Action": Institutional Analysis of the Piracy Dilemma in China

Roscoe Pound, the founder of the movement for "sociological jurisprudence", made a distinction between "law in books" and "law in action" in his famous article:[247]

> [I]f we look closely, distinctions between law in the books and law in action, between the rules that purport to govern the relations of man and those that in fact govern them, will appear, and it will be found that today also the distinction between legal theory and judicial administration is often a very real and very deep one.[248]

Similarly, Karl Llewellyn, a prominent legal realism scholar, has emphasized the discrepancy between the "paper rules" and "real rules".[249] The essential enlightenment of these dichotomies is to remind us that it is important to look at what the legal system actually is rather than what legislators or judges say it is. The insight underlying this distinction indicates that the focus of observing law must be also upon the law in action, not merely the law on the books. This may be viewed as a call for deeper research into legal sources, historical backdrops and the social context that surround legal rules.

This book borrows Roscoe Pound's distinction and highlights the gap between the statements of statutory law and the actual implementation of the law. It is not my purpose here to engage with all the controversies concerning the relationship between the law on the books and the law in action. My target, more modestly, is to identify and analyze the deep-seated causes of the contradiction between the superficial high protection in law and the ineffective low protection actually offered by the law in action.

In the last two decades of 20 century, the Chinese copyright legislative progress underwent unprecedented development against the backdrop of China's

---

247 Roscoe Pound, *Law in Books and Law in Action*, 44 AM. L. REV. 12 (1910).

248 *Id*, at 15.

249 See Karl N. Llewellyn, *A Realistic Jurisprudence – The Next Step*, 30 COLUM. L. REV. 431, 447-54 (1930) (describing and distinguishing "paper rules" and "real rules" and explaining their application to doctrine); see also KARL N. LLEWELLYN, JURISPRUDENCE: REALISM IN THEORY AND PRACTICE 24 (The University of Chicago Press 1962).

entering the WTO and constructing a law-based society. The phenomenon of copyright infringement remains severe in practice despite the improvement of the legal system. In comparison with the implementation of other laws, one may ask why IP law has been confronted with such difficulty.

There is an apparent difference between Western countries and China in the process of constructing IP laws. Over the course of hundreds of years in the West, the progress of IP law development has largely been in synch with economic development in each country. Yet, the current speed of IP legislation in China obviously exceeds its rate of economic growth. What are the causes behind this institutional transition? What are the factors derogating from the effectiveness of IP law implementation and self-conscious adherence to the law?

We should consider the things which must be ensured or satisfied and not simply consider the system charged with ensuring or satisfying them, as if the system's existence was its own ultimate goal.[250] It is indeed necessary to establish a sound copyright system to protect individual's intellectual products in the literary field. However, the justification would be suspect if we only cared about the existence of the system itself.

In respect of the rampant copyright infringements taking place in China, this can be seen as a result of comprehensive institutional causes rather than the mere imperfection of the law. Arguably, the indigenization of a foreign legislative system, the limitations of the copyright implementation "dual system", the absence of a notion of protecting private rights, as well as the high pricing of legal copies in the market constitute the realistic causes of the piracy dilemma in China. In this section, I will analyze these underpinning factors from the perspectives of law making, law implementation, adherence to the law, and the market.[251]

2.2.1 Foreign Pressure and China's Copyright Regime: A Review of Chinese Copyright Law Making History

The copyright system is considered exotic in China. The first neoteric Chinese copyright law – the 1910 Copyright Act – was formulated at gunpoint, and was

250 Roscoe Pound, Falü shi jieshi [INTERPRETATION OF LEGAL HISTORY] 226 (Beijing: Zhongguo fangzheng chubanshe [China Fangzheng Press] 2002).

251 A complete view of "regulation" should consider four modalities, including the law, social norms, the market and architecture. Each constraint constitutes a "regulator". Changes in any regulator will affect the regulation of the whole. Some constraints will support others; some may undermine others. See LAWRENCE LESSIG, CODE VERSION 2.0 123 (NY: Basic books 2006).

not implemented. In the initial period of the PRC, copyright policies experienced multiple twists and turns. Beginning with the 3rd plenary session of the 11th central committee of the CPC in 1978, the foundation of a socialist Chinese copyright law started to spread.

2.2.1.1 Law at Gunpoint: A History of the Neoteric Chinese Copyright Law

It seems quite surprising that the copyright system remained almost completely invisible in ancient China, a country where paper making and printing – the two main technologies of knowledge transmission – had been invented. "Since all primary conditions of the industrial revolution in 18th century Britain are recognized by economists and historians to have been almost ready in 14th century China,"[252] why did China fail to promulgate the first intellectual property law in the world?

On the basis of investigation into the character of the law, the status of esquires, and the spirit of merchants, some Chinese scholars argue that in Imperial China authors shared the idea of being "a transmitter rather than a maker, believing in and loving the ancients",[253] or the idea that "one who excels in study can follow an official career".[254] These beliefs or traditions provided a substitute tool for the archaic clerisy to claim the rights to their works. In addition, there was no demand for rights allocation since merchants widely relied on the bureaucracy.[255]

In spite of the existence of a prohibition against piracy promulgated by the government in the Song Dynasty (960 – 1276),[256] these phenomena were regarded as a means of controlling ideology by the Chinese emperor, rather than

---

252 See Lin Yifu, Zhidu, jishu yu zhongguo nongye fazhan [SYSTEMS, TECHNOLOGY AND THE AGRICULTURE DEVELOPMENT IN CHINA] 244 (Shanghai: Shanghai San lian shudian, Shanghai renmin chubanshe [Shanghai San Lian Bookstore & Shanghai Renmin Publishing House] 1994).

253 See the Analects of Confucius, chapter 7, Shu er pian (1), Shu er buzuo, xin er haogu.

254 A Confucian slogan for education, see the Analects of Confucius, Zi Zhang pian, Shi er you ze xue, xue er you ze shi [Zi Xia said, "The officer, having discharged all his duties, should devote his leisure to learning. The student, having completed his learning, should apply himself to be an officer."].

255 See Yi Yufeng, Sixiang kongzhi yu quanli baohu [*Mind Control or Right Protection*], Xinan zhengfa daxue boshi lunwen [Ph.D. Dissertation, Southwest University of Political Science and Law], chapter 2, pp.59-73 (2003).

256 See Zhongguo banquan shi yanjiu wenxian [LITERATURE OF THE CHINESE COPYRIGHT HISTORY] 3 (Zhou Lin & Li Mingshan eds., Beijing: Zhongguo fangzheng chubanshe [China Fangzheng Press] 1999).

the protection of individual's property rights. In the long history of Chinese feudal dynasties, when "Kings had long arms, and all people were king's men", there was no conception of private intellectual property rights or civil liability.[257]

In fact, the primary laws protecting intellectual property rights including copyright were passed under the pressure of Western countries.[258] In the late 19th century, foreign traders possessed a series of privileges in China. In order to escape from the high outback tax, or to obtain transit privileges in the backlands, some Chinese traders counterfeited certain foreign trademarks. For example, in 1897 a British trader sued an opium processor in Shantou in respect of the unauthorized use of its opium brand. Foreign traders' IPRs could not, however, receive sufficient protection on account of the lack of an IP system in China, while at the same time they could not invoke the IP laws in their home countries directly. In order to clear obstacles to entering and occupying the tremendous Chinese market, Westerners felt it was necessary to urge China to establish an IPR legal system modeled on the Western framework.[259]

In order to obtain the agreement of the Qing government, Westerners made a compromise that if China instituted IP laws, they would agree with the Chinese custom yamun to renew impost and forbid opium. If the status of legislation and its implementation were guaranteed, they would even give up extraterritoriality. The promise of respecting China's sovereignty and protecting the state's safety attracted the Qing government's interests. Moreover, such conditions were also contained in the Peace Treaty of 1901.

During negotiations, the issue with which the Chinese delegates were most unacquainted was IPR protection. Cultural diversity induced fierce debates among the delegates. For instance, the Chinese side thought that copyright protection would raise the price of literature beyond the ability of poor people to purchase it. In light of this negative impact on the enlightenment of the citizenry, copyright was seen as incompatible with the Western desire to urge China to enter world civilization. On the other hand, the Western side insisted that without copyright protection, the right holders of excellent foreign works would hesitate to enter China. This would block the modernization steps of the

---

257 See An Shoulian, Zhishi chanquan haishi sixiang kongzhi: dui zhongguo gudai fa de wenhua toushi [*Intellectual Proptery or Thought Control: A Cultural Interpretation of Chinese Acient Law*] in Falü de wenhua jieshi [A CULTURAL INTERPRETATION OF LAW] (Liang Zhiping ed., Shanghai: San Lian shudian [San Lian Bookstore] 1998).

258 See Zhang Yumin & Li Yufeng, Zhongguo banquan shigang [*A Historical Outline of the Chinese Copyright System*], 1 Keji yu falü [SCIENCE-TECHNOLOGY AND LAW] 42, 44 (2004).

259 *Id.*

Qing government.[260] As there was little diplomacy for weak countries, in the end, the Qing government promised to protect the foreign works and constitute a copyright law.

The Qing government promulgated the first copyright law in neoteric China, the 1910 Copyright Act.[261] However, it was too late to put it into force as the coming revolution in the next year ended the history of the Qing dynasty. In the succeeding decades, there were two copyright laws in the neoteric Chinese history. They were the 1915 Copyright Law promulgated by the Northern Warlords Government and the 1928 Copyright Law enacted by the National Government.[262] These two copyright laws also failed to be implemented during the chaos of wartime.

In a word, the ideal of copyright protection and its legal framework in neoteric China unfolded an exotic character. The gap between the notion of western IPRs and the Chinese tradition to some extent slowed down the process of institutional indigenization. Together with lack of effective implementation, copyright protection was deeply rooted neither at the national level nor in the heart of the Chinese population.

2.2.1.2 Ups and Downs: Copyright Policy of the PRC from 1949-1976

After the founding of the PRC, the aforementioned three neoteric copyright laws were abolished as symbols of corrupt and degenerate feudal ideology. The Chinese government began to build its new socialist copyright mechanism in light of model provided by the Soviet Union. Remuneration was protected in order to encourage creativity by intellectuals.

In 1950, the first national publication meeting held by the Ministry of Culture promulgated the Resolution on the Improvement and Development of Publication Work (hereinafter the Publication Resolution). It emphasized the need for publishers to respect copyright and publishing rights, and that they must not engage in reprinting, copying or tampering with works or other similar

260 See *id.* For the history of the process of the 1910 Copyright Act promulgated by the Qing dynasty, see generally Wang Lanping, Jindai zhongguo zhuzuoquan fa de chengzhang (1903-1910) [DEVELOPMENT OF THE COPYRIGHT LAW IN NEOTERIC CHINA: 1903-1910] (Beijing: Beijing daxue chubanshe [Peking University Press] 2006).

261 For more details of the 1910 Copyright Act, see Zhou Lin & Li Mingshan eds., *supra* note 256, 89-94.

262 For the comparison of the three copyright laws in neoteric Chinese history, see Wang Lanping, Zhongguo fazhi jindaihua guocheng zhong de sanbu zhuzuoquan fa [*Three Copyright Laws in the Process of Modernization of Chinese Law*], 3 Bijiao fa yanjiu [JOURNAL OF COMPARATIVE LAW] 44 (2005).

misbehavior. The key issue of this meeting was the payment of remuneration. It was agreed that the remuneration should be negotiated with the author according to the principle of respecting the interests of authors', readers' and publishers'. For the author's sake, it was forbidden to completely sell their copyright in principle.[263]

Although the Publication Resolution in essence was not law but an administrative regulation, and the author's right only limited to obtaining remuneration, it played a very important role before the establishment of the copyright law. The Publication Resolution altered the ambiguous attitude towards the author's interest, and it gradually became the basis for settling copyright disputes. An initial case pursuant to the Publication Resolution was a publication dispute between Xinhua Bookstore Dalian branch and the World Knowledge Press. Dalian branch pirated a book – *International Situation after the Korea War* – published by the World Knowledge Press. The latter appealed in light of the Publication Resolution to the Publication Headquarters of the Central People's Government which was in charge of national publication at that time. As a result, Xinhua Bookstore made a self-criticism and compensated the loss of World Knowledge Press.[264]

Until 1955, the social status of intellectuals had gradually risen during the process of rural collectivization and industrialization. "The new environment and situation requires intellectuals, especially those senior intellectuals possessing academic positions, to make more contribution to the socialist construction."[265] However, this harmonious situation was broken by the "Hu Feng event" in 1955.[266] After that, the Communist Party confronted a passive intelligentsia who evaded work, development and research.[267]

In order to seek the cooperation of intellectuals in economic development, on the 2nd of May, 1956, Chairman Mao advocated the famous policy of "letting a hundred flowers blossom and a hundred schools of thought contend". The

263 See Shen Rengan & Zhong Yingke, Banquan fa qianshuo [ELEMENTARY INTRODUCTION OF THE COPYRIGHT LAW] 106 (Beijing: Falü chubanshe [Law Press China] 1982).

264 See *id.*

265 See Guangming ri bao [GUANGMING DAILY], Dec. 3, 1955.

266 See Zhang Xiaofeng, Wo yu Hu Feng: Hu Feng shijian sanshiqi ren huiyi [HU FENG AND I: IN MEMORY OF THE HU FENG EVENT BY THIRTY SEVEN PEOPLE] (Yinchuan: Ningxia renmin chubanshe [Ningxia People's Publishing House] 1993); Mei Zhi, Hu Feng zhuan [*Biography of Hu Feng*], in Mei Zhi wen ji [THE WRITING OF MEI ZHI] (Xiao Feng ed., vol. 3, Yinchuan: Ningxia renmin chubanshe [Ningxia People's Publishing House] 2007).

267 R. MacFarquhar & Fei Zhengqing eds., Jianqiao Zhonghua Renmin Gongheguo shi [THE CAMBRIDGE HISTORY OF CHINA] 255 (Beijing: Zhongguo shehui kexue chubanshe [China Social Sciences Publishing House] 1998).

policy encouraged independent thinking and free debate, and further stimulated intellectuals' enthusiasm and courage to call for institutional reform and the establishment of a copyright law.[268]

In response, the Ministry of Culture promulgated a draft of Provisional Regulations of Ensuring Publication Copyright (Baozhang chubanwu zhuzuoquan zanxing guiding (cao'an)). The draft which included 11 articles mainly used the Soviet Union's Basic Regulations of Copyright as a reference. It granted copyright to literary works, orally delivered works, translations, music scores, artistic drawings, scientific graphics, and maps. In addition, it provided the ownership of copyright and the transfer of copyright. It also included the copyright of certain derivative works, and the relief of copyright infringement.

Unfortunately, the draft was not enacted. After June 1957, the "Anti-rightist Movement", aiming to eradicate attacks against the Party, and the "Great Leap Forward Movement", aiming to quicken the pace of socialist development, involved politics becoming the center of all tasks. Besides, the political validity of intellectuals who were engaging in creativities was suspected during these two movements.

As a reflection of the political tendency in that historical stage, the Chinese government not only hesitated to establish IP laws, but also deprived the chance of intellectuals to receive high monetary incentives. In the copyright field, in July 1958, the Ministry of Culture issued a draft of Provisional Regulations on the Remunerations of Literary and Social Science Works (Guanyu wenxue he shehui kexue gaochou de zanxing tiaoli cao'an) in order to unify the inconsistent remuneration payed by different publishers.[269] The purpose of the draft was "on the one hand, to prevent the paying of remuneration too low which would go against the incentive for further creation and the improvement of creativity level; on the other hand, to avoid the paying of remuneration too high which would baffle or impair author's further creativity by virtue of inducing the tendency of pursuing creature comfort, despising labor, looking down on workers and peasants, and deviating politics and reality." [270]

---

268 See Guanyu "Baozhang chubanwu zhuzuoquan zanxing guiding (caoan) de shuoming" [Instruction Concerning the Draft of Provisional Regulations of Ensuring Publication Copyright].

269 Wenhuabu [Ministry of Culture], Ban fa "Guanyu wenxue he shehui kexue shuji gaochou de zanxing guiding caoan," qing Beijing, Shanghai liangdi youguan chubanshe shixing [Issue Concerning the Draft of Provisional Regulations on the Remunerations of Literary and Social Science Works, and Piloting in the Relative Publishing Houses in Beijing and Shanghai], July 14, 1958.

270 Yi Yufeng, *supra* note 255, at 140.

However, the "Great Leap Forward Movement" was promoted further after the Beidai River Conference passed the decision establishing the People's Commune in August 1958. China started to carry out the policy of "large in size and collective in nature" ("Yi da er gong"), and the supply system. Under the great enthusiasm of the communist movement, from September 1958, a number of authors and publishers began to advocate reducing remuneration. Later, the Ministry of Culture held a symposium and invited more than thirty organizations, including People Daily, Guangming Daily, People's Press, and the Chinese Writers Association.

The participants reached a consensus that the current remuneration standard was too high to adapt to the new situation of the "Great Leap Forward Movement". The excessive high remuneration caused the life specialization of a portion of intellectuals, and this separated them from the masses of workers and peasants. It deepened the artificial discrepancy between brainwork and physical work which was unsuitable to the rising communist consciousness. Moreover, some intellectuals did not concentrate on their work and studies, but instead pursued remuneration which propagated capitalist individualism. Therefore, the consensus was to reduce by half of the then current remuneration in all newspapers and magazines.[271]

Subsequently, in order to implement the consensus of the symposium and ensure the appropriate standard of authors' living, the Ministry of Culture issued the Instructions on Several Problems Concerning the Reduction of Remuneration Standard (Guanyu jiangdi gaochou biaozhun de jige wenti de tongzhi) in March 1959, and the Instructions on the Publishers in Beijing and Shanghai Districts to Continue the Tentative Implementation of the "Provisional Regulations on the Remuneration of Literary and Social Scientific Books" (Guanyu Beijing, Shanghai liangdi youguan chubanshe jixu shixing "guanyu wenxue he shehui kexue shuji gaochou de zanxing guiding" de tongzhi) in October 1959. Moreover, during the climax stage of the "Great Leap Forward Movement", the Ministry of Culture and the Chinese Writers Association even composed a proposal to fundamentally reform the remuneration system and abolish copyright royalty. At the same time, they suggested implementing a wage system for the writers who entirely lived on remuneration, and to make

271 Wenhuabu [Ministry of Culture], Guanyu Beijing ge baokan, chubanshe jiangdi gaochou biaozhun de tongzhi [Information Concerning the Reduction of the Payment of Author's Remuneration by Newspapers, Periodicals and Publishing Houses in Beijing], Oct. 10, 1958.

remuneration only exist as a subsidy of life and a necessary factor of promoting creativity.[272]

Economic confusion during the late 1950s and early 1960s compelled the CCP to modify the People's Commune system. In order to rectify the extremely low remuneration in the period of the "Anti-rightist Movement" and the "Great Leap Forward Movement", and to recover the 1959 Provisional Regulations on the Remuneration of Literary and Social Scientific Books, in August 1961, the Ministry of Culture issued the Instructions Concerning Appropriate Implementation of the Remuneration Standard (Guanyu zhengque zhixing gaochou zhidu, qiadang zhangwo gaochou biaozhun de tongzhi). The Instructions established the principle that "one who does more work will get more pay" and opposing the equalitarianism as a means of promoting author's enthusiasm and their contribution to the prosperity of socialist literature and art project.

However, the policy of protecting intellectuals' remuneration was once again denied during the national "Rectification Movement" launched by CCP in 1964 against the capitalist ideology. As a result, the Ministry of Culture put forward the suggestion to further reduce the remuneration of publications in newspapers, periodicals, and books.[273]

The intellectuals' interests were further derogated during the ten years of "Cultural Revolution" from 1966 to 1976. Not only was the then current copyright system seriously damaged, but also large numbers of intellectuals were "purged". The order of politics, economy, and literature was in great chaos. Everything was covered by the color of "revolution". Rich and colorful dramas were substituted by the sole templet drama. Writers, scientists, and artists were derided as the "stinking ninth category"[274] and sent to the country or work farms to receive physical force training.[275]

---

272 See Wenhuabu dangzu, Zhongguo zuojia xiehui dangzu guanyu feichu banshuizhi, chedi gaige gaochou zhidu de qingshi baogao [A Report Submitted by the Leading Party Group of the Ministry of Culture, and the Chinese Writers Association Concerning the Abolishment of Copyright Royalty and the Revolution of Author's Remuneration], 1960.

273 See Wenhuabu [Ministry of Culture], Guanyu jinyibu jiangdi baokan shuji gaochou de qingshi baogao [A Report Concerning Further Reduction of the Payment of Remuneration by Newspapers, Periodicals and Books], Jan. 3, 1966.

274 "Chou lao jiu" was a term abused by ultra-leftists for teachers, writers, artists, and other educated people in the 1966-1976 Cultural Revolution. They were regarded as class enemies next to landlords, reactionaries and even spies.

275 See Fei Zhengqing, Weida de Zhongguo geming [THE GREAT CHINESE REVOLUTION 1800-1985] (Beijing: Shijie zhishi chubanshe [World Affairs Press] 2000).

At the same time, not only did authors lose the protection of their moral rights, but also even the reduced remuneration system was abolished.[276] Nobody had the courage to admit the role of creativity.[277] Even the identity of the author was suspected. Just as a popular saying in the "Cultural Revolution" stated, "Is it necessary for the steel workers to engrave their names on their productive steel? If not, why do the intellectuals enjoy the privilege to sign their names on works?"[278]

2.2.1.3 The Foundation of Socialist Copyright Law: Requirements of the Opening-up policy and the Normalization of China-US Relationship

After the 3rd plenary session of the 11th central committee of the CPC in 1978, the "take the class struggle as the outline" policy in the "Cultural Revolution" was abolished. The construction of democracy and a legal system in China started to recover and develop. A series of measures aimed at promoting creativity by individuals were carried out, especially the reinstatement of the college entrance examination and the reconstruction of the IPR legal framework.

In October 1977, the National Publication Bureau (Guojia chuban ju) issued the Notice Concerning the Tentative Implementation of the Remuneration and Subsidy of Press and Publication (Guanyu shixing xinwen chuban gaochou ji butie banfa de tongzhi). Similarly, the Ministry of Culture promulgated the Provisional Regulation on the Remuneration of Books (Guanyu shuji gaochou de zanxing guiding) and reestablished the remuneration system of 1958.

Notably, the establishment and improvement of the Chinese copyright system were greatly influenced by the legal systems in the US and other Western countries. After signing two important bilateral conventions – the US-China high-energy physics cooperation and the US-China Trade Agreement – in 1979, China promised to protect copyright and other IPRs.[279]

In April 1979, a report concerning the drafting of a copyright law and gradually entering the international copyright conventions was submitted to Mr. Hu Yaobang, the director of national propaganda. He replied promptly: "The

276 See Shen Rengan & Zhong Yingke, *supra* note 263, at 108.

277 A report of XINHUA DAILY in Oct. 1966 wrote that, in the Daqing Oilfield, no one applied for the remuneration of significant discoveries or inventions, and no one applied for a patent.

278 Yi Yufeng, *supra* note 255, at 129.

279 See Article 6 of the US-China high-energy physics cooperation (Zhongmei gaoneng wuli xieding); Article 6 of the US-China Trade Agreement (Zhongmei maoyi guanxi xieding).

report is approved. Please hurry up to organize a team and draft the copyright law." This report was regarded as the beginning of the lawmaking process of the contemporary Chinese copyright protection.[280] From that time forward, IPRs began to be protected as private rights.[281] Before that, although the PRC had promulgated some Regulations, Measures, or Instructions concerning copyright, the private ownerships of authors to their creative works and their property rights to receive remuneration had never been admitted.

The making of copyright law is full of twists and turns, as is the fate of Chinese intellectuals.[282] As one scholar observed, the copyright law was one of the most controversial and time-consuming laws ever approved by the National People's Congress (NPC).[283] The dilemmas between private rights and national interests, and between creative diversity and social stability, were the focus of disputes during the making of the 1990 Chinese Copyright Law.

Different protests represent diverse group interests and political positions. As Wang Hanbin, the vice-director of the NPC indicated, the copyright law was the most complicated one in the legislative history of the PRC, having gone through more than 20 drafts.[284] Regretably, the first 1990 Chinese Copyright

280 Zhou Lin, Zhongguo banquan shi yanjiu de jitiao xiansuo [*Some Clues for Researching Chinese Copyright History*], in Zhongguo banquan shi yanjiu wenxian [LITERATURE OF CHINESE COPYRIGHT HISTORY] (Zhou Lin & Li Mingshan eds., Beijing: Zhongguo fangzheng chubanshe [China Fangzheng Press] 1999).

281 Zheng Chengsi, Zhishi chanquan baohu shiwu quanshu [A PRACTICAL MANUAL OF INTELLECTUAL PROPERTY PROTECTION], foreword (Beijing: Zhongguo yanshi chubanshe [China Yanshi Press] 1995).

282 Tan Hongkai, *New Copyright Law Is a Welcome Start*, CHINA DAILY, Oct. 12, 1990.

283 Jiang Ping, Jiang Ping wenji [SELECTED WORKS OF JIANG PING] 71 (Beijing: Zhongguo fazhi chubanshe [China Legal Publishing House] 2000).

284 There were several competing arguments. For instance, one suggested that the Chinese copyright law should accord with the international standard in order to introduce foreign culture and technologies into China. Although the process might be painful, it would promote the Chinese literature and technology market from a long-term perspective. A counter argument stood by the national interest and social stability perspective and considered that given the fact that the lawmaking of copyright was ahead of the press law, it should bear the temporary function of the latter. That was to say, the Chinese copyright law should adhere to a socialist character, and serve for the socialist economic base and superstructure. Any works forbidden by the law did not have copyright. Another opinion suggested that the Chinese copyright law should adapt to the the primary stage of socialism, and the step of reform and opening-up. See Shen Rengan, Banquan fa yu banquan baohu de shehui zuoyong [*Social Function of Copyright Law and Copyrihgt Protection*], in Banquan yanjiu wenxuan [SELECTED WORKS OF COPYRIGHT RESEARCH] (Zhongguo banquan yanjiuhui [Copyright Society of China] ed., Beijing: Shangwu yinshuguan [The Commercial Press] 1995); Huang Shuhai et al,

Law was a compromise of contradictions which failed to grant rights as propagandizing at home and abroad.[285] An obvious example was the Chinese remuneration system. At that moment, it was regarded as a limit of the author's economic rights. The Chinese government had set up a maximum standard for the remuneration of literary works, translations, dictionaries, and other items.[286] Even if the work possessed very high value in the market, it could not receive a copyright royalty higher than the prescribed price.[287]

Meanwhile, the Chinese copyright law bore the functions of the press law's political review and ideological control for reasons of maintaining social stability.[288] The legitimacy requirement of copyrighted works reflected the demands of the socialist spiritual and material civilization, as well as the social public order.[289] In response, Western commentators observed, ideology was a consideration of the Chinese copyright law. For instance, it was prohibited to spread the works which would harm national unity, sovereignty and territorial integrity; instigate national hatred and separatism; disturb the social order, and promote other illegal behavior against the "Four Basic Principles".[290] For those works which were worthless to national stability, the National Press and

---

Guanyu woguo zhuzuoquan lifa de jidian sikao [*Considerations on Chinese Copyright Lawmaking*], in *id*; Xiao Xun, Cong woguo de shiji chufa jiejue banquan (zhuzuoquan) lifa zhong you zhengyi de wenti [*Working Solution for Disputed Issues in Copyright Lawmaking*], in *id*; Zheng Chengsi, Xinxi chanquan yu banquan fa [*Information Property Right and Copyright Law*], in *id.*

285 See Peter Schloss, *China's Long-Awaited Copyright Law*, 5 CHINA BUSINESS REVIEW 24-28 (1990).

286 For instance, the Maximum Remuneration per thousand words is 40 yuan for literature, 35 yuan for translation, and 30 yuan for ancient books. See Shuji gaochou zanxing guiding [Provisional Regulations Governing Remuneration for Book Manuscripts], issued by NCAC on June 15, 1990.

287 See Liu Gushu, *Question of Worldwide Interest in Connection with the Chinese Copyright Law*, 4 CHINA PATENT & TRADEMARK 21-23 (1991).

288 Article 4 of the 1990 Chinese Copyright Law provides that: "Works the publication or distribution of which is prohibited by law shall not be protected by this law. Copyright owners, in exercising their copyright, shall not violate the Constitution or laws or prejudice the public interests."

289 See Zhang Yumin ed., Zhishi chanquan faxue [INTELLECTUAL PROPERTY JURISPRUDENCE] 70 (Beijing: Zhongguo jiancha chubanshe [China Procuratorial Press] 2002).

290 In March 1979, Deng Xiaoping introduced the "Four Basic Principles", including the principles that "we must keep to the socialist road; we must uphold the dictatorship of the proletariat; we must uphold the leadership of the Communist Party; and we must uphold Marxism-Leninism-Mao Zedong Thought."

Publication Bureau controlled their transmission either through direct censorship or indirect rejection of book numbers.[291]

In spite of several limitations, the first 1990 Chinese Copyright Law has actively exerted significant functions in promoting literature and the arts, and in protecting the private rights of individuals. It not only created the basic framework of modern copyright law, but also admitted author's private interests as one of its legislative goals.

Song Muwen, the director of the NCAC at that time highly evaluated the law. He said that the implementation of the copyright law had a positive effect in protecting copyright holders' legitimate interests, providing incentives for intellectuals to create, and promoting the development of the economy and science, and the prosperity of culture and art. Along with the implementation of the copyright law, the copyright administration, judicial trials, theoretical research, organization building, international communication and cooperation had also made great progress. Citizen's protection consciousness of copyright was also strengthened to some extent.[292]

In order to enter the WTO, it was necessary to establish consistency between the Chinese IP laws and TRIPS. However, the making of the 1990 Chinese Copyright Law considered more of the current Chinese situation, and left a gap with international treaties.[293] Soon after the enactment of the 1990 Chinese Copyright law, great political, economic and diplomatic pressures pushed China to join in several international treaties and bilateral agreements with the US. Discrepancies induced double standards that led to foreigners receiving higher levels of protection than Chinese.[294]

---

291 See WILLIAM P. ALFORD, TO STEAL A BOOK IS AN ELEGANT OFFENSE 79 (California: Stanford University Press 1995).

292 See Song Muwen, Guanyu woguo zhuzuoquan fa de xiugai [*Concerning the Revision of the Chinese Copyright Law*], in 1 Chuban kexue [PUBLISHING SCIENCE] 4 (2002).

293 When constituting the 1990 Chinese Copyright Law, there was no intent to join the Berne Convention and other international treaties within a year or two. Thus, some provisions remained in obvious conflict with the Berne Convention. For example, the restriction of the right of reproduction, performance, and broadcasting exceeded the permitted scope of the Convention.

294 For instance, Article 43 of the 1990 Chinese Copyright Law provides that a radio station or television station may broadcast for noncommercial purposes a published sound recording without seeking permission from or paying remuneration to the copyright owner. In order to implement the Berne Convention, the State Council promulgated Provisionas on the Implementation of the International Copyright Treaties (Promulgated on Sep. 25, 1992 by Decree No. 105 of the State Council of the People's Republic of China, and effective as of Sep. 30, 1992). The Provision only solved the protection for

Apparently, such double standards would derogate from China's political image. More importantly, after two decades of economic reform, the consciousness of the market economy had been formed, and democratization had been gradually advanced through political reform. Especially under the background of "rule of law", individuals had increasingly strengthened their consciousness of private rights, and began to actively claim for their rights.[295]

In July 2001, the Amendment Act of the 1990 Chinese Copyright Law was passed. There were three main modifications: first, reducing or removing the gap between the Chinese copyright law and international treaties, especially the Berne Convention, the TRIPS, as well as bilateral agreements to abolish double standards; second, solving practical problems in implementing the copyright law, including clarifying the status and functions of the copyright collective society, strengthening the enforcement of copyright law, and prescribing the transfer of property rights; thirdly, solving the urgent legal problems raised by the emergence of new technology, such as databases, multimedia and software.

In summary, tracing the history of the process of establishing the law, the Chinese copyright law has the exotic characteristics of a legal transplant. From the first neoteric Chinese copyright law drafted at gunpoint, to the ups and downs of copyright policy in the PRC, as well as the difficult foundation of a socialist copyright law, the twists and turns remind us that the effective implementation and self-conscious adherence to the copyright law could not be accomplished in one action.

2.2.2 Limitations of the Copyright Implementation "Dual System"

In China, administrative agencies have played an important role since the establishment of the copyright regime, and together with the courts formed the so-called "dual system". This bipolar mechanism has played an important role in the process of establishing a legal regime in a very limited period and in protecting copyright against rampant piracy, counterfeits and other infringements. However, on the negative side, administrative protection

foreigners pursuant to international treaties and it, however, still retained the application of the current domestic copyright law to the Chinese.

295 Quanguo Renda Changweihui Fazhi Gongzuo Weiyuanhui Minfashi [Cilvil Law Office, Law Committee of the Standing Committee of the National People's Congress] ed., Zhonghua Renmin Gongheguo zhuzuoquan fa xiugai lifa ziliao xuan [SELECTED LEGISLATIVE MATERIALS OF THE CHINESE COPYRIGHT LAW] (Beijing: Falü chubanshe [Law Press China] 2002).

currently dominates this dualism and induces multiple problems, which to some extent, derogate from the effect of copyright protection and anti-piracy efforts.

In this part, I will describe the status quo of the administrative dominant copyright "dual system" and illustrate the limitations of an administrative dominant copyright protection regime – principally, agency overlap and minoritarian bias – from both normative and descriptive perspectives.

2.2.2.1 The Administrative Dominant "Dual System"

In this section I will explore the current situation of the copyright dual protection system. Notably, at present administrative protection actually dominates the bipolar mechanism. I will describe this characteristic in two tracks: first the powerful administrative protection and second the correspondingly weak judicial protection.

2.2.2.1.1 Powerful Administrative Protection

Copyright Competent Authority

At present, administrative domination is the primary characteristic of the copyright dual protection system. Chinese administrative agencies have played the leading role in copyright protection. According to statistics, in all social realms including IPRs, about 80% of legislative regulations are implemented by administrative agencies.[296] IPRs administrative agencies possess comprehensive and powerful authority. Take for example the competent authority in charge of copyright – the NCAC. The following diagram (table 2.1) makes a comparison

296 See Ying Songnian, Yifa xingzheng lungang [*On the Administration by Law*], 1 Zhongguo faxue [CHINESE LEGAL SCIENCE] 29, 33 (1997).

of the functions of the US Copyright Office,[297] Japan Copyright Office (JCO)[298] and the NCAC.[299]

Table 2.1: Comparison of the Functions of Copyright Administrations

Administration	US Copyright Office	JCO	NCAC
Registration	With Copyright Office; optional; but prerequisite of infringement action with respect to U.S. works only; registration pre-infringement requisite to statutory damages and attorneys' fees to all works	With JCO; optional; presumption of authorship, first publication, date of creation and being effective against any third party on the matters of transfer, alteration, expiry or the restriction on the disposal, of the right of pledge established on copyright	With CPCC; optional; *prima facie* evidence of copyright ownership
Administrating compulsory and statutory	Administers compulsory and statutory licenses	Administers compulsory and statutory licenses, but	Administers statutory licenses

297 Available at http://www.copyright.gov/circs/circ1a.html (last visited Sep. 8, 2010).

298 Notably, most countries of civil law tradition basically do not have an independent copyright administration. For example, in Germany, the Federal Ministry of Justice administers copyright affairs, while in France, the Ministry of Culture and Francophone Affairs, Directorate of General Administration, Under-Directorate of Legal Affairs and Office of Literacy and Artistic Property take charge of copyright issues. Similarly, JCO (Chosakukenka) is not an independent government administration but belongs to the Agency for Cultural Affairs (Bunkachō, ACA), which is a part of the Ministry of Education, Culture, Sports, Science and Technology (Monbukagakushō, MEXT). For more details about JCO and ACA, available at http://www.bunka.go.jp/chosakuken/index.html, and also http://www.cric.or.jp/cric_e/csj/csj.html (last visited Sep. 10, 2010).

299 For more details about the functions of the NCAC, available at http://www.ncac.gov.cn/GalaxyPortal/inner/bqj/include/detail.jsp?articleid=4618&boardpid=869&boardid=1150 1010111613 (last visited Sep. 10, 2010).

licenses		scarcely implementing	
Legislation	Provides technical assistance to the Congress and to executive branch agencies; establishes administrative regulations	Planning of copyright legislations	Participates in the drafting of copyright law and regulations; Constituting copyright administrative regulations
Enforcement		Countermeasures against piracy from policy perspective	Investigates and punishes copyright infringement cases

a. Registration

Since Article 5 (2) of the Berne Convention confirmed the automatic vesting of copyright by forbidding the imposition of formal prerequisites to enjoy and exercise copyright or its enforcement on foreign Berne member works, most countries, including Japan [300] and China, [301] have also ceased to require copyright-constitutive formalities, mainly the notice of copyright, registration, deposit and recordation, to domestic authors as preconditions of copyright warranty and protection as well.

Although registration is voluntary and optional, given its significant effect in facilitating the ascertainment of copyright status and the ownership of works, domestic copyright laws customarily encourage registration. For instance, the Copyright Law of Japan explicitly lists the act of registration on the presumption of authorship (Article 75 (3)), first publication (Article 76), date of creation (Article 76 *bis*) and being effective against any third party on the matters of

300 Article 17(2) of the Copyright Law of Japan provides: "The enjoyment of moral rights of authors and copyright shall not be subject to any formality."

301 Article 2 of Proposed Measures of Works Voluntary Registration provides: "The registration of works is voluntary. The enjoyment and the exercise of copyright shall not be subject to registration."

transfer, alteration, expiry or the restriction on the disposal, of the right of pledge established on copyright (Article 77).[302]

Similarly, although the Chinese copyright law does not explicitly stipulate registration matters, various regulations recognize the effect of registration as *prima facie* evidence of copyright ownership.[303] Registration is especially encouraged in the field of software copyright, software exclusive licensing contracts and assignment contracts.[304] Notably, as part of the undergoing administrative agency reform, the original registration organ – the NCAC has transferred its registration affairs to the Copyright Protection Center of China (CPCC).[305]

By contrast, the US has adopted a "two-tier" registration system[306] that relaxes formal requirements on foreign works in light of the Berne Convention,[307] but retains registration as a prerequisite for an infringement action with respect to US works.[308] In addition to this pre-suit requirement, the

302 The matters also shall not be effective against any third party without the registration of the right of publication. See Article 88 of the Copyright Law of Japan.

303 Article 1 of Proposed Measures of Works Voluntary Registration provides: "The registration of works is beneficial to clarify the copyright ownership and provide primary evidence for the right holder in the case of infringement." See also Interpretation of Applicable Law on Copyright Civil Disputes by the Supreme Court which treats copyright registration certificate as primary proof; and Measures for the Implementation of the Regulation on the Customs Protection of IPRs by the General Administration of Customs which regards works registration certificate as the required document of IPRs customs protection recordation.

304 Article 2 of Registration of Computer Software Copyright Procedures provides: "the copyright administrative department of the State encourages registration of software and gives priority protection to registered software." See also its Article 3.

305 The CPCC is a social copyright management and social service organization under the direct leadership of the General Administration of Press and Publication of the PRC (PPA) and the NCAC. It is responsible for the registration of copyright in computer software, registration of pledge contracts on copyright, registration of copyright in various published works, verification and registration of publishing contracts of imported audio-video products, and public information. For more details, available at http://www.ccopyright.com.cn/cpcc/index_en.jsp (last visited Sep. 12, 2010).

306 See Jane Ginsburg, *The US Experience with Formalities: A Love/Hate Relationship*, For ALAI Congress, London, June 17, 2009, p.21, available at http://www.alai2009.org/Presentations/Jane%20Ginsburg%20US%20Experience%20with%20formalities%20(ALAI%202009)%20rev.doc (last visited Sep. 12, 2010).

307 See Article 5 (3) of the Berne Convention. It provides that the "[p]rotection in the country of origin is governed by domestic law". Accordingly, it is permitted for the US to maintain the requirement on the formalities with respect to the US works.

308 See 17 U.S.C. §411 (a).

US copyright law also encourages the registration of all works regardless of origin by confirming registration as *prima facie* proof of validity of copyright[309] and the prerequisite of statutory damages and attorney's fees.[310]

b. Administration of compulsory license and statutory license

Compulsory licenses and statutory licenses are limitations on a copyright holder's exclusive rights. Under these the licensee is required to pay royalties to the copyright holder even though they did not obtain permission from the holder. In addition to explicitly itemizing statutory circumstances that do not require permission, copyright compulsory licenses are justified as a means of moderating the danger that exclusive licenses can be used to create market power in downstream markets.[311]

The copyright competent authority is responsible for the administration of compulsory licenses and statutory licenses under copyright law. The US copyright law provides in detail for copyright statutory licenses [312] and compulsory licenses. [313] The broad application of involuntary licenses is regarded as representing a shift of the copyright system from a property regime to a liability regime.[314] Analogously, the Copyright Law of Japan itemizes the applicable circumstances for copyright statutory licenses [315] and compulsory licenses.[316] However, in practice the compulsory license is criticized for being rarely implemented. [317] In contrast, Chinese copyright law does not include

309 See *id.* §410 (c).

310 See *id.* §412.

311 See Alan Fisch, *Compulsory Licensing of Blacked-Out Professional Team Sporting Event Telecasts (PTSETS): Using Copyright Law to Mitigate Monopolistic Behavior*, 32 HARV. J. ON LEGIS. 403, 417-28 (1995); Michael Meuer, *Vertical Restraints and Intellectual Property Law: Beyond Antitrust*, 87 MINN. L. REV. 1871, 1906 (2003).

312 See 17 U.S.C. §112, §114, and President Signs into Law the Small Webcaster Settlement Act of 2002 (SWSA).

313 See 17 U.S.C. §115.

314 See Daniel Crane, *Intellectual Liability*, 88 TEX. L. REV. 253 (2009).

315 See Articles 30 (2), 33, 33 *bis*, 34, 36 and 38 of the Copyright Law of Japan.

316 See *id.* Articles 67-70. For overviews of the literature, see Nobuhiro Nakayama, Chosakukenhō [COPYRIGHT LAW] 331-34 (Yuhikaku 2007); Yoshiyuki Tamura, Chosakukenhō gaisetsu [COPYRIGHT LAW] 513-17 (2d. ed. Yuhikaku 2006); Fumio Sakka, Shōkai chosakukenhō [INTERPRETATION OF COPYRIGHT LAW] 405-10 (3rd. ed. Gyosei 2006).

317 Nakayama observes that now the compulsory license system is seldom used. E.g., Article 67 "Exploitation of works in the case where the copyright owner thereof is unknown" is in fact hardly applied; Article 68 "Broadcasting of works" is only used in

compulsory licenses, and restricts statutory licenses to only a few limited instances.[318]

c. Legislation

In contrast with the US Copyright Office and the JCO just providing suggestions to the legislators, the NCAC can actively participate in the drafting of copyright law.

Arguably, the whole process of copyright legislation and revision mirrors the significant role of the NCAC. In the wake of the founding of the PRC, the NCAC established the Drafting Committee on Copyright Provisional Regulation and submitted the draft in November 1957, although discussions on it were terminated by the Cultural Revolution. After the 3rd Plenary Session of the 11th CPC Central Committee in December 1978, the copyright legislation again entered the schedule. This time the direct impetus was the China-US Trade Agreement which required China to protect copyright. In response, on April 21, 1979, the NCAC submitted a report concerning the copyright issues in the China-US Trade Agreement to the State Council, which suggested immediately selecting personnel, constituting a special committee, collecting materials, engaging in investigation and training personnel, and assisting in the drafting of copyright law.

The same story was repeated during the modification of the copyright law with the coming of the new millennium. In January 1998, the NCAC put forward the basic revision principles and submitted an explanation concerning the modification of the copyright law to the State Council. The NCAC not only drafted the copyright law and its modification proposal, but also created the implementing regulations of copyright law[319] as well as various copyright administrative regulations.

In addition to providing technical assistance to the legislator,[320] the US Copyright Office can also establish administrative regulations.[321] The JCO, however, does not have the authority to institute administrative regulations.

---

exceptional occasions; and nowadays the existing significance of Article 69 "recording on commercial phonograms" is very slim. See Nakayama, *id.* at 331-34.

318 See Articles 23, 33(2), 40(3), 43(2) and 44 of the 2010 Chinese Copyright Law.

319 Article 54 of the 1990 Chinese Copyright Law indicates that: "The implementing regulations of this Law shall be drawn up by the copyright administration department under the State Council and shall enter into force after approval by the State Council."

320 See 17 U.S.C. §701 (b) (1), (2).

321 See *id.* §702.

d. Enforcement

The US Copyright Office and the JCO principally do not have authority over copyright enforcement. Although the JCO manages countermeasures against piracy, this is mainly from a policy perspective.[322]

By contrast, the NCAC has the authority to investigate and punish copyright infringement cases. Article 48 of the 2010 Chinese Copyright Law authorizes the copyright administration to take charge of copyright infringement cases if it affects the public interest. The enforcement measures of the copyright administration principally include putting a stop to the infringing act, confiscating unlawful income from the act, and charging administrative penalties.

In a word, notwithstanding the fact that the authority of the NCAC in respect of registration and involuntary licenses is comparatively weaker, its mandatory mission in the drafting of copyright law and enforcement are apparently distinct. Buried beneath this powerful influence in the legislative process as well as in enforcement, the legislature is exposed to the risk of being "captured" by the administrative agencies and the danger of policy stagnation or minoritarian bias increases as a result.

Special Offices

Besides the independent administrative agencies, special offices, notably the National Office of Rectification and Standardization of Market Economic Order,[323] and the National Anti-Pornography and Anti-Illegal Publications Office,[324] have been established to organize and harmonize the combined enforcement of various administrative agencies.

To take one example, the leadership and influence of the National Anti-Pornography and Anti-Illegal Publications Office in mobilizing and uniting dispersed administrative forces can hardly be reached by any separate administration. Table 2.2 illustrates the main functions of the office and the member units which cover various ministries in culture, education, finance,

---

322 See Kaizokubantaisaku ni tsuite [*Countermeasures against Piracy*], available at http://www.bunka.go.jp/chosakuken/kaizokuban/gaiyou.html (last visited Sep. 15, 2010).

323 For more details on the National Office of Rectification and Standardization of Market Economic Order, available at http://law.baidu.com/pages/chinalawinfo/4/77/469222793f404a524ca28e63184010e4_0.html (last visited Sep. 9, 2010).

324 Zhongguo sao huang da fei wang [website of China Anti-Pornography and Anti-Illegal], available at http://www.shdf.gov.cn/portal/index.html (last visited Sep. 8, 2010).

commercial, transport, judicial as well as police areas. In addition to its headquarters in Beijing, there are branches in 31 provinces and cities which cooperate with the local administrative agencies.[325]

Table 2.2: Functions and Member Units of the National Anti-Pornography and Anti-Illegal Publications Office

Function[326]	Participates in the rule-making of publication market and the policy of Anti-Pornography and Anti-Illegal Publications
	Administers the publication process and market
	Investigates illegitimate publishing behaviors and sequestrates illegal publications.
	Organizes, coordinates and instructs collective action or special governance of "Anti-Pornography and Anti-Illegal Publications" among various administrations
	Supervises and urges important cases
Member Units[327]	Publicity Department of the CPC Central Committee Commission of politics and law of the CPC Central Committee General Office of the State Council National Copyright Administration The Supreme People's Court The Supreme People's Procuratorate Ministry of Culture Ministry of Education Ministry of Industry and Information Technology Ministry of Public Security Ministry of Supervision

325 For more details about the branches, available at http://www.shdf.gov.cn/portal/org_developement.html (last visited Sep. 8, 2010).

326 For more details about the functions of the National Anti-Pornography and Anti-Illegal Publications Office, available at http://www.shdf.gov.cn/portal/org_function.html (last visited Sep. 10, 2010).

327 Available at http://www.shdf.gov.cn/portal/org_member.html (last visited Sep. 8, 2010).

	Ministry of Civil Affairs Ministry of Finance Ministry of Transport State Administration for Industry & Commerce State Administration of Radio Film and Television State Post Bureau General Administration of Customs General Administration of Press and Publication Civil Aviation Administration The Publicity Department of People's Liberation Army The Political Department of People's Armed Police Forces …

Through combined enforcement and special actions, Chinese administrations have obtained great achievements in copyright protection. For instance, in 2008 China captured 83,837,000 illegal publications, banned 46,000 shops and 8 illegal CD product lines, ceased 1420 printers and 14,000 illegal websites, deleted 490,000 detrimental pieces of network information, investigated and dealt with 25,056 administrative punishment cases and 328 criminal cases. The statistics on seized illegal publications which decreased for the first time in recent years revealed that various illegal publishing activities were being effectively stifled.[328]

Thus, it would not be an exaggeration to say that administrative protection plays a dominant role in copyright protection, notably in the areas of copyright infringement enforcement. The protection of copyright and other IPRs is not only the responsibility of the competent authorities, but also an important task of all Chinese government agencies.[329]

328 2008 nian "sao huang da fei" gongzuo qude xianzhu chengji [*In 2008 the task of "Anti-Pornography and Anti-Illegal Publications" obtained great achievements*], available at http://www.gapp.gov.cn/cms/html/47/711/200902/462578.html (last visited Sep. 8, 2010).

329 In recent years, all of the annual Reports on the Work of the Government refer to the protection of intellectual property.

2.2.2.1.2 Weak Judicial Protection

The original purpose of the "dual system" was to overcome the disadvantage of the single approach and obtain mutual complementation. Article 7 (2) of the Chinese Administrative Punishment Law states explicitly the relationship between administrative enforcement and criminal justice: "[w]here offences against the law constitute crime, investigation shall be conducted to determine criminal responsibility; administrative punishments are not to replace criminal punishments".

However, in practice administrative protection dominates the copyright "dual system" and impairs judicial protection. Considerable infringement cases which are supposed to be delivered to the court and charged with criminal liability are now being substituted with administrative penalties. This is on account of the lack of efficient connections and information sharing platforms, due process requirements[330] as well as necessary supervision of administrative behavior.

To take one example, in 2006, the Shanghai Administration for Industry & Commerce investigated 5189 counterfeit and piracy cases with the total value involved of 830,000,000 RMB, among which 78 cases were major cases valued at more than 1,000,000 RMB, while only 21 cases were handed over to the judiciary. The Shanghai Bureau of Quality and Technical Supervision inquired into 1685 infringement cases, while only 15 cases were delivered to the court. The Shanghai Administration of Customs examined 2148 smuggling cases, captured 132 suspects with the total value involved of 113,400,000,000RMB, while only 41 cases and 81 suspects received decisions from the court.[331]

One utilitarian reason underlying this phenomenon is that the administrative penalty supplements the financial resources of the administration beyond the national financial allocation, and is thus closely related to the welfare of

330 The administrative agencies in China conventionally highlight substantive justice but overlook procedural justice. The attitude that the administrative process would restrict the efficiency of administrative protection is actually incorrect. "The nature of the process is to prevent corruption, arbitrariness and exorbitance in the administration and decision making." See Ji Weidong, Fazhi zhixu de jiangou [CONSTRUCTION OF LEGAL ORDER] 57 (Beijing: Zhongguo zhengfa daxue chubanshe [China University of Political Science and Law Press] 1999).

331 See Cai Junyong, Jidai jianli xingzheng zhifa yu xingshi sifa xianjie jizhi [*A Junction System between Administrative Enforcement and Criminal Justice Is Urgently Needed*], Jiancha ribao [PROCURATORIAL DAILY] Sep. 20, 2009, available at http://www.jcrb.com/n1/jcrb1423/ca638707.htm (last visited Sep. 8, 2010).

personnel.[332] The scope of enforcement authority has potential effects on office conditions, personnel income, labor insurance and welfare in administrative agencies.

The weakness of copyright judicial protection is the important underpinning of the existence of administrative dominance. In contrast with powerful administrative protection of copyright, judicial protection in China is rather weak. Although it has undergone decades of development, the IPRs judicial protection has improved and has obtained remarkable achievements.[333] Yet in the present stage, judicial protection is still too feeble to bear the task of IPRs protection alone. Concretely speaking, the following aspects impair the effects of judicial protection.

To begin with, there is no unified special judicial organ to deal with copyright cases. Copyright civil, administrative and criminal cases are dispersed into different tribunals. Moreover, their diverse operations – where copyright first instance civil cases are dealt with mainly by intermediate courts while copyright first instance administrative and criminal cases are still dealt with by primary courts – induce conflicts of verdicts. In practice, there were several cases where the criminal tribunal of primary courts initially rendered a guilty verdict, whereas the civil tribunal of intermediate courts found no infringement.[334] In these circumstances, the latter courts enter into a dilemma: if they render verdicts of no infringement, their decision would conflict with

---

332 For instance, a few administrative agencies hook enforcement with the economical interests of personnel, and treat administrative penalties as the way to earn extra revenue. The Legislative Committee of Sichuan People's Congress suggests that government of all levels should not distribute administrative penalty indexes or hook the penalty with welfare and allowance of enforcement personnel. More details are available at http://news.sina.com.cn/c/2002-01-18/448589.html (last visited Sep. 8, 2010).

333 According to statistics, in 2005 the local courts all together dealt with 3567 IPRs infringement cases, a year-on-year growth of 28.36%; 3529 cases reached a verdict, year-on-year rise of 28.28%. Among which 524 were IPRs criminal cases, year-on-year growth of 35.40%; 505 cases were ended, 741 suspects were charged with criminal liability, which indicated that criminal judicial protection of IPRs is being strengthened. See Zuigao Renmin Fayuan Zhishi Chanquan Ting [Intellectual Property Tribunal of Supreme Court], 2005 nian Zhongguo zhishi chanquan sifa baohu gaikuang [*Survey on 2005 Chinese IP Judicial Protection*], 4 Falü shiyong [JOURNAL OF LAW APPLICATION] 7 (2006).

334 Chen Huizhen & Xu Jun, Lun woguo zhishi chanquan liti shenpan moshi de goujian [*On the Construction of Chinese IP Solid Trials*], 4 Falü shiyong [JOURNAL OF LAW APPLICATION] 12, 12-13 (2006).

effective criminal judgments; whereas if they maintained the infringement verdicts, two misjudged cases would be engendered.[335]

In addition to this, the implementation standard is not unified. On account of the imbalance of economic and social development among different districts, the quality and quantity of copyright cases diverge dramatically. In general, most copyright cases are dealt with by the courts in developed eastern coastal regions while few are heard in the Midwest. The IPRs cases in some primary courts of developed districts even equal or exceed the total cases in all courts of certain undeveloped provinces. The lack of cases in undeveloped areas stagnates the training of judges and the accumulation of experience to handle specific IPRs cases, especially new types of infringements. The discrepancy in the judge's quality induces the discrimination of verdicts towards similar cases in different courts and influences the uniform implementation of copyright law.

Admittedly, heavy reliance on administrative agencies may be a rational choice in the initial stages of establishing a copyright regime. This is particularly so when adapting copyright to an unsound legal system with limited judicial resources and a dim defense consciousness on the part of copyright holders. Since adopting the "reform and opening up" policy in 1978, China has been in the process of large scale commercial and social development and innovation. The multiple governance resources in China, involving custom, stipulation, clan rules and other "local resources",[336] along with powerful governance by the Party and administrations, bear the weight of social governance together with the law.[337] During this specific period, the domination of government power has had the dramatic comparative institutional advantage of unifying planning, centralizing forces and mobilizing all possible social resources. It would not be an exaggeration to say that the significant achievements of establishing a copyright regime within the past 25 years, and the provision of some degree of protection for copyright holders' interest has, to a great extent, depended on the administrative dominant approach.

---

335 Zuigao Renmin Fayuan Minsan Ting [The Third Civil tribunal of Supreme Court], Zhishi chanquan shenpan zhidao yu cankao [IP TRIAL GUIDANCE AND REFERENCE] 138 (vol.7, Beijing: Falü chubanshe [Law Press China] 2004).

336 See Su Li, Fazhi jiqi bentu ziyuan [THE NATIVE RESOURCE OF RULE OF LAW] 3-22 (Beijing: Zhongguo zhengfa daxue chubanshe [China University of Political Science and Law Press] 1996).

337 These governance resources, however, on the other hand limit the operation of law, and make it difficult to completely achieve the principle of "supremacy of law" in the current stage. See Gu Peidong, Ye lun Zhongguo faxue xiang hechu qu [*Where Will China's Legal Science Go?*], 1 Zhongguo faxue [CHINA LEGAL SCIENCE] 5, 14 (2009).

In contrast, pursuant to public interest theory, the other institutions might bear some critical comment. Principally, the legislature typically lacks specialized information and expertise; courts are regarded as decentralized and lack expertise; and markets are deemed decentralized and uncontrolled. Thus, administrative agencies are more likely to achieve the objective of bringing complex phenomena under control than any of these rival institutions.[338]

In this part, I have dwelt on the administrative dominant copyright protection regime. I now turn to an analysis of its main limitations, utilizing some concepts of institutionalism.

2.2.2.2 Limitations of an Administrative Dominant Copyright Protection Regime

Notwithstanding the fact that reliance on an administrative approach was a realistic and rational choice in the initial stages, over time the "public interest" focus of the administrative agencies has given way to "capture pathology". In China, the decentralized model of IPRs administration induces agency overlap and policy conflict. The competition among multi-headed administrative agencies either runs into stagnation and generates policy inconsistency, or represents "minoritarian bias" when one agency captures the legislature.

According to Richard Stewart, agencies are often accused of subverting public interests in favor of the private interests of regulated and client firms.[339] The capture hypothesis suggests that once administrative agencies are given substantial authority to make and enforce policy decisions, they are likely to be subjected to extensive pressure from groups that have a particularly strong interest in the consequences of their policy determinations. The focus of this pressure would invariably be an attempt to promote the private interest of the regulated group at the expense of some broader interest of the public.[340]

In this section, I will not cover all flaws of the administrative dominant copyright protection regime, but only pick up three particular limitations that are potentially relevant to an international audience. I will explore how these concepts of institutionalism, notably Komesar' theory, are illuminated in a distinct Chinese context.

---

338 See Merrill, *supra* note 52, at 1049.

339 See Richard Stewart, *The Reformation of American Administrative Law*, 88 HARV. L. REV. 1669, 1682-83 (1975).

340 See Mark Seidenfeld, *A Civic Republican Justification for the Bureaucratic State*, 105 HARV. L. REV. 1511, 1565-70 (1992).

2.2.2.2.1 Agency Overlap: Multi-headed Copyright Administration

Overlapping organizations and functions, and conflicting policies from different departments are long-standing and well-documented problems within administrative systems in China. Since the 1978 start of the "reform and opening up", China has already undergone five administrative reforms.[341] At present, the sixth administrative reform is underway with a view to building a service-oriented government as indicated in the Report to the Seventeenth National Congress of the CPC on October 15, 2007.[342] It is focused on streamlining government organs, downsizing various organs for deliberation and coordination, cutting down levels of administration, minimizing costs, and establishing greater departments with integrated functions.

In the IPRs field, the problems of agency overlap and policy conflict also exist, and are compared in a lively way to a "multi-headed" administration.[343] In contrast with most countries in the world that have adopted centralized IPRs administrations, China has chosen a decentralized model.[344] The current IPRs

---

341 The focuses of these five administrative reforms are: in 1982 the abolishment of life tenure in leading posts; in 1988 "transfer of government function"; in 1993 "adapting to the requirements of establishing socialism market economy system"; in 1998, the cancellation of industrial economy special departments to overcome the fusion of government and corporation; in 2003, administrative system renovation. See Xinhua, Jiedu gaige kaifang yilai wuci zhongyao de zhengfu jigou gaige [*Interpretation on the Five Governmental Institution Reforms Since the Beginning of the Reform Process and Opening Up*], available at http://news.xinhuanet.com/politics/2008-01/28/content_7509409.htm; more details about Chinese administrative reforms see Xinhua News Special Topic, Shenhua xingzheng guanli tizhi he jigou gaige [*Deep Administrative System and Institutional Reform*], available at http://www.xinhuanet.com/politics/xzgltzgg/ (last visited Jan. 6, 2010).

342 The sixth administrative reform will establish several new ministries including those in the areas of industry and information, transportation, human resources and social security, environment protection as well as housing and urban-rural development. This administrative reform has not referred to IPRs administrations.

343 See e.g., Li Chi, Xingzheng guanli guocheng zhong duotou zhifa de jiejue lujing tanxi [*On the Solutions to Multi-headed Enforcement in the Administrative Process*], 7 Fazhi yu shehui [LEGAL SYSTEM AND SOCIETY] 198 (2009); Fang Jun, Dangqian xingzheng guanli zhong duotou zhifa, duoceng zhifa wenti de sikao yu duice [*Countermeasures to Deal with Multi-headed Administration*], Beida falü xinxi wang [website of Law Information China], available at http://article.chinalawinfo.com/Article_Detail.asp?ArticleID=37444 (last visited Feb. 2, 2010).

344 At present, among 196 countries and areas which implement an IP regime, there are 180 adopting a centralized IP office. However, only less than 10 have chosen a decentralized administration. See Zhishi chanquan guanli gailun [INTELLECTUAL PROPERTY

administration framework is mainly based on the separation of authority to classify the IPRs protection into 10 departments.[345] Although the Patent Office was renamed the State Intellectual Property Office (SIPO) in 1998, it did not change the decentralized arrangement of IPRs administration. The SIPO inherited former patent administrative affairs and only added the function of overall planning and coordination.

As regarding copyright affairs, the NCAC is the competent authority in charge of copyright, along with the Ministry of Culture, the State Administration for Industry & Commerce (SAIC), the Ministry of Commerce, and the General Administration of Customs. Arguably, the enormous institutions and elaborate division of administrative authority have not promoted more efficient administration and effective protection, but conversely induced overlap, inefficiency, and inconsistency among agencies.

Overlap and Inefficiency

A copyright administrative protection system with diffused and overlapping lines of authority and responsibility cannot efficiently accomplish its mission and meet new protection challenges. The decentralized arrangement of copyright administration seems on the surface to reach every aspect, however, in reality each administrative agency is always trying to expand the jurisdictional lines which generate the authority scramble in profitable areas whilst striving to shrink accountability and result in "buck-passing" in nonprofit areas. Agencies are, consequently, criticized as unthinking "bureaucracies" more concerned with

---

MANAGEMENT CONSPECTUS] 35 (Luo Guoxuan ed., Zhishi chanquan chubanshe [Intellectual Property Press] 2007).

345 The State Intellectual Property Office takes charge of patent and layout-designs of integrated circuits; The NCAC is responsible for copyright issues; The Trademark Office of State Administration for Industry & Commerce deals with trademark issue; The Antimonopoly and Anti-unfair Competition Enforcement Bureau of State Administration for Industry & Commerce presides over antimonopoly and anti-unfair competition affairs; The General Administration of Quality Supervision, Inspection and Quarantine deals with geographical indications; The Ministry of Agriculture and State Forestry Administration separately handle new varieties of agricultural plants and forestry plants; The Ministry of Commerce tackles IPRs affairs in international trade; The Ministry of Science and Technology takes charge of IPRs relevant to science and technology; and the General Administration of Customs handles IPRs relevant to inward and outward goods.

expanding their budgets than with attending to the needs of the beneficiaries of regulation.[346]

Perhaps one notable shortcoming in the current copyright administrative system is its arbitrary jurisdictional lines. These lines cause agency overlap and administrative inefficiency. Given copyright administration and protection are regulated by more than one agency, multiple government representatives, each with varying guidelines, have to undertake multiple inspections of the same object.

Notwithstanding that the NCAC is the competent authority, given that the characteristics of copyright works relate to literature, culture and art, other agencies also have jurisdiction to intervene. For instance, some copyright works belong to the field of culture and arts, and it is the authority of the Ministry of Culture to stipulate principles and policies for the development of culture and arts in China, to draft laws and regulations on culture and arts, to administer literary and artistic undertakings, and guide artistic creation and production.[347] When the copyrighted works enter the market and are threatened by piracy, monopoly and unfair competition, or other illegal acts which damage market order and the legitimate rights and interests of right holders and consumers, they belong to the mandate commission of the SAIC.[348] When related to domestic copyright and foreign trade as well as international cooperation, these issues are under the supervision of the Ministry of Commerce.[349] When involving inward and outward copyright goods, it involves the monitoring of the General Administration of Customs.[350]

The copyright administrative jurisdiction is split among several largely uncoordinated agencies, increasing the risk of duplicative efforts and administrative inefficiency. The multiple administrative departments each have their own organization system, office space and personnel, and devote a lot of resources to dealing with similar missions which greatly waste public

346 See WILLIAM NISKANEN, BUREAUCRACY AND REPRESENTATIVE GOVERNMENT 81 (Chicago: Aldine, Atherton 1971).

347 For more details about the function of Ministry of Culture, available at http://www.ccnt.gov.cn/English/Introduction/200904/t20090428_62706.html (last visited Mar. 8, 2010).

348 For more details about the function of the SAIC, available at http://www.saic.gov.cn/english/aboutus/Mission/ (last visited Mar. 8, 2010).

349 For more details about the mandate of the Ministry of Commerce, available at http://english.mofcom.gov.cn/mission.shtml (last visited Mar. 8, 2010).

350 For more information about the mission of the General Administration of Customs, available at http://english.customs.gov.cn/publish/portal191/ (last visited Mar. 9, 2010).

resources.[351] Additionally, as a result of agency overlap, one or more departments tend to neglect their responsibilities in reliance on other agencies.[352]

Administrative inefficiency also happens between copyright and other IPRs protection. In recent years, the IPRs cases have tended to be more complicated. Yet the existence of jurisdiction lines blocks the collaboration of settlements among different categories of IPRs. For example, if the suspected copyright case which the NCAC is investigating also involves the trademark or patent infringement, the NCAC can only deal with the copyright issue and the whole case can not be completely solved until the Trademark Office of the SAIC and the SIPO participate. The dispersed jurisdiction delays the dispute investigation, increases the protection cost, and decreases the settlement's efficiency. On account of this, during the IPRs bulletin in Shanghai to 24 consulates, the issue which most concerned foreign parties was which one agency – as opposed to several – should be referred to in the case of an IPR dispute.[353]

Notwithstanding the fact that the combined enforcement of various administrative agencies have been undertaken on the national and local scales, the legitimate basis of combined enforcement itself is questionable. If various agencies possess the same authority, it is undoubtedly a redundant set up and a waste of resources. If these agencies lack relevant authority or only part of these agencies have jurisdiction, then can unauthorized organs be empowered for reasonable purposes in combined enforcements? If that is the case, does it breach the fundamental principle of "statutory authority"? Even if we do not take account of the legitimacy of their source of power and solely evaluate based on the effectiveness of combined enforcement, this still does not fully meet expectations. Usually after the "movement", illegal actions are repeatedly revived.[354] As the IPRs infringements become more and more complicated, it is imperative to streamline agencies and explore ways to establish greater departments with integrated functions.

---

351 Zhu Xuezhong & Huang Jing, Shilun woguo zhishi chanquan xingzheng guanli jigou de yitihua shezhi [*On the Unified IP Administrative Agency in China*], 3 Keji yu falü [SCIENCE TECHNOLOGY AND LAW] 82 (2004).

352 Wang Yingming, Shichang jingji ying zhuzhong dui zhishi chanquan de xingzheng baohu [*The Market Economy Needs More Emphasis on the Administrative Protection of IPRs*], 11 Xingzheng yu fa [PUBLIC ADMINISTRATION & LAW] 78, 80 (2008).

353 See Zhu Xuezhong & Huang Jing, *supra* note 351, at 83.

354 Liu Wenjing, *supra* note 215, at 73.

Policy Inconsistency among Agencies

The decentralized and dispersed system of administration leads to diverse protection standards and conflicting policies among rival departments when the competition plunges into stagnation and fails to reach a compromise.

This has been the story concerning the charging of copyright royalty fees on Karaoke. Since 2003 the ubiquitous Karaoke service has increasingly triggered disputes between copyright holders and bar operators for copyright infringement in China.[355] At the present these private disputes have been upgraded into a conflict of departmental interests. In 2006, two Chinese administrative agencies came up with separate plans to charge karaoke bars copyright royalty fees. This caused a turf war. On July 18, the Ministry of Culture said it would create the "National Karaoke Management and Service System" which allows karaoke bars to voluntarily connect to the system without expense, but charges them for each song downloaded. The purpose of the Ministry of Culture in creating the unified karaoke system is to block public access to "unhealthy songs" and stop the infringement of copyright. The system contains a list of songs that are copyright authorized, and it can record how many times a song is ordered, indicating how much copyright-holders should be paid.[356]

Later on July 20, the NCAC announced its own karaoke copyright royalty fees system and on November 9, the NCAC unveiled a pilot scheme that required karaoke bars in Beijing, Shanghai and Guangzhou to pay 12 yuan (1.5 US dollars) a day for each private room (less in underdeveloped regions). Two national associations – the Music Copyright Society of China (MCSC) and the China Audio-Video Collective Administration (CAVCA) – were entrusted with the fee collection.[357]

The two administrative agencies had not, however, conferred with each other about the fees, and as yet no moves seem to have been made to unify the charge system. At a press conference, the NCAC spokesman Wang Ziqiang stated that the Ministry of Culture was not entitled to charge karaoke operators

---

355 Fifteen domestic and foreign recording companies, including Warner Music, Global Music and Sony BMG, launched a campaign in 2003 to charge royalties from Chinese Karaoke operators who had lost a succession of lawsuits and had been fined thousands of yuan. See Xinhua, *Two Chinese Gov't Bodies Jockey for position in Karaoke Market*, available at http://news.xinhuanet.com/english/2006-08/04/content_4918539.htm (last visited Sep. 8, 2010).

356 See Xinhua, *China to Unify KTV System to Protect IPR*, available at http://news.xinhuanet.com/english/2006-07/19/content_4855063.htm (last visited Sep. 8, 2010).

357 See Xinhua, *China's Karaoke Bar Royalty Scheme Reaches Impasse*, available at http://news.xinhuanet.com/english/2006-11/29/content_5408694.htm (last visited Sep. 8, 2010).

because copyright supervision and fee collection was the responsibility of the NCAC. While Liang Gang, an official of the Ministry of Culture, said to the China Newsweek magazine that the Ministry had established a platform for trade between copyright holders and bar operators, and that it had authority over the collective pricing and supervision of song content.[358] Although the two administrative agencies' moves are both said to be aimed at protecting right holder's interests, copyright holders were excluded from the royalty pricing process.

In fact, the separate proposals with distinct charge mechanisms actually reflect a tussle of interests between the NCAC and the Ministry of Culture. Collecting copyright royalty fees for the use of musical and video products is said to be a potentially lucrative business. According to a survey, China has an estimated 100,000 karaoke establishments – each with an average of 10 private rooms – generating almost 1 billion yuan in turnover annually.[359] The Collective Management of Copyright Regulations that came into effect on March 1, 2005, allows copyright collective management bodies to take a certain percentage from the royalties.[360] Wang Huapeng, deputy director of the CAVCA, said that the association would receive a cut – less than 20 percent – of the collected royalties to cover its management cost. According to Wang's calculation, the MCSC, with 4,234 members, collected 64 million yuan (8 million US dollars) in royalties in 2005, which would net the association more than 12 million yuan (1.5 million US dollars).[361] To some extent, the two government agencies are competing for the power to collect those sums.

"We should pay the royalty charges but the problem is to whom and how we should pay," Huang Shiqiu, head of the Guangzhou Cultural and Entertainment Industry Association, which represents the city's largest karaoke bars, said to Xinhua.[362] The unified karaoke system of the Ministry of Culture seems more humane, but it is voluntary; whereas in the NCAC scheme, those which refuse to pay will be deemed illegal. Yet, the per-room charge system of the NCAC suffered opposition from karaoke bars arguing that the two entrusted associations with fee collection actually did not have the legal authorities to do so and the charges were unreasonably high. Wang Xudong, a copyright lawyer in Nanjing, said that "The two associations can only collect royalties for

---

358 See Xinhua, *supra* note 355.

359 See Xinhua, *China Closes First Karaoke Bar over Copyright Infringement*, available at http://www.chinadaily.com.cn/bizchina/2009-12/24/content_9225932.htm (last visited Sep. 8, 2010).

360 See Article 28 (1) of the Collective Management of Copyright Regulations.

361 See Xinhua, *supra* note 357.

362 *Id.*

copyright owners who have entrusted them to do so. In fact, the copyright of most songs played in karaoke bars has not been entrusted to these associations."[363] He added that since copyright is a private right, royalty charges should be negotiated by the copyright owners and the bar operators rather than established by the administrative agencies.[364] Meanwhile, karaoke consumers are worried that even if operators accept the fee due to administrative pressure, in the end they would transfer the fees to consumers.[365]

In this part, I have explored the pathology of agency overlap. The multi-headed copyright administration induces overlap and inefficiency. When the conflicting measures among rival agencies plunge into stagnation, policy inconsistencies are generated. However, when one agency captures the legislature, another different pathology – minoritarian bias appears.

2.2.2.2.2 Minoritarian Bias: Agency Interest in Copyright Legislative Process

Komesar has presented a two-force model of malfunctions in the political process, which he has labeled majoritarian bias and minoritarian bias.[366] In the case of majoritarian malfunction, a tyrannical majority can victimize a discrete and insular minority.[367] The second form of malfunction, minoritarian bias, indicates that concentrated high-stake interests often prevail over dispersed small-stake interests in the legislative process as well as in public agency decision-making.[368] "From the standpoint of resource allocation efficiency, minoritarian bias occurs when a concentrated high per capita minority prevails over the dormant low per capita majority even though the total social costs imposed on the losing majority are greater than the total social benefits gained by the successful minority."[369]

In the Western context, minoritarian bias has been utilized in myriad contexts in the literature on rent-seeking and interest group politics.[370] Yet in

---

363 *Id.*

364 *Id.*

365 Liang Xingguo, Kala OK banquan shoufei zhi zheng yu zhishi chanquan zhong de gongquan wenti [*Public Rights Issues of Intellectual Property in the Disputes of KTV*], 1 Shanghai caijing daxue xuebao [JOURNAL OF SHANGHAI UNIVERSITY OF FINANCE AND ECONOMIC] 46, 52 (2008).

366 See KOMESAR, IMPERFECT ALTERNATIVES, *supra* note 51, at 53-8, 213-21.

367 See *United States v. Carolene Products Co.*, 304 U.S. 144, 152, n.4 (1938).

368 See Engelbrekt, *supra* note 173, at 68.

369 See KOMESAR, IMPERFECT ALTERNATIVES, *supra* note 51, at 76.

370 See James E. & Ruth B. Doyle-Bascom, *Book Review: A Comment on Imperfect Alternatives: Choosing Institutions in Law, Economics, and Public*, 83 CALIF. L. REV. 941, 946 (1995).

China it is embodied through departmental interest in legislation. In contrast to the conventional agency capture theory which states that different interest groups fight to capture one dominant agency to influence legislation, in China the battle is among multi-headed administrative agencies to capture the legislature.

Departmental protection is an unusual phenomenon in the Chinese legislative process, which arises along with the participation and intervention of administrative agencies in legislation. A tradition has formed that whenever the NPC and its Standing Committee exercise the legislative power of the State to constitute or revise any law, other corresponding administrative agencies are invited to discuss and solve the main problems or key issues, while incidentally giving birth to the notion of departmental protection. Departmental protection has been unfolding on two main tracks. The first one is directed at expanding the authority of the department, while the second one, seeking to obtain various interests, notably the economic benefits that come with maintaining or enlarging departmental authority.[371]

This has been the story of copyright law, to give the most representative example. The long, time consuming, redundant and tortuous process of legislation can hardly be exceeded by other laws. It took 11 years of repeated consultation and modification, and had to be submitted to the Standing Committee of the NPC 4 times for deliberation before the draft was finally passed in 1990. 11 years later, the copyright law was revised in order to meet the requirements of entering the WTO and to adapt to the new situation. This time it also took 3 years for reiterative amendments and 3 deliberations by the Standing Committee of the NPC in contrast to the normal dual examinations. No wonder some scholars commented that a similar instance was scarcely encountered in legislative history before.[372]

One significant reason underlying the extraordinarily troublesome creation and revision of the copyright law was that it involved multiple departmental interests which were difficult to harmonize and balance. To take one representative example, the former Article 43[373] was regarded by many

371 See Liu Songshan, Guojia lifa sanshi nian de huigu yu zhanwang [*Reflection and Prospects of China's Legislation in the Past Thirty Years*], 1 Zhongguo faxue [CHINA LEGAL SCIENCE] 31, 40 (2009).

372 Zheng Bochao, Zhuzuoquan fa zai zhengbian zhong qian xing, xiugai wenti zaici bei tiqi [*The Disputes on the Revision of the Chinese Copyright Law*], Jiancha ribao [PROCURATORIAL DAILY], Sep. 1, 2009, available at http://newspaper.jcrb.com/page/1/2009-09/01/04/2009090104_pdf.pdf (last visited Feb. 8, 2010).

373 Article 43 of the 1990 Chinese Copyright Law indicates that a radio station or television station may broadcast, for noncommercial purposes, a published sound recording

legislators as the most difficult obstacle in copyright legislation and modification,[374] because it touched off the conflicting interests between copyright holders and intermediary business organizations – namely, radio stations and television stations.

In fact, in early 1989, the copyright draft submitted by the NCAC regulated the use of radio stations and television stations as statutory licenses. Without seeking permission from copyright holders, stations still had to pay remuneration to them. The NCAC's statutory license suggestion was obviously in favor of the copyright holder's interest. However, when the draft was deliberated on by the Standing Committee of the NPC, notwithstanding support by the musicians and record industries, the approach of "statutory license" encountered fierce opposition from the State Administration of Radio Film and Television (SARFT). The primary reason of the SARFT was that the radio and television stations in China were state owned and noncommercial organizations which bore an important socialist propaganda mission to promote public welfare. The operation funds were appropriated by the government. If the legislation required the payment of remuneration, the Ministry of Finance would have to pay the bill.[375] The objection of the SARFT finally "captured" the legislature and the former Article 43 was passed.

The triumph of the SARFT, on the one hand, owed much to the bigger picture of a planned economy that caused policy to customarily favor state-owned organizations. On the other hand, the case illuminated (borrowing the terminology of Komesar) a minoritarian bias in the legislative process. Interest group theory argues that agencies are likely to become "captured" by the groups that they are charged with regulating. The disproportionate influence these groups wield over administrative agencies influences the legislature.[376] Institutions seek to maximize their own self-interested ends in the way they respond to these interest groups.[377] This seems to be an accurate description of what happened in China in the case of Article 43.

Pursuant to the capture assumption, the concentrated high-stake interests of stations were likely to prevail over the dispersed small-stake interests of

---

without seeking permission from, or paying remuneration to, the copyright owner, performer and producer of the sound recording.

374 See the comment by one legislator – Song Muwen, *supra* note 372.

375 See *id.*

376 See William Landers & Richard Posner, *The Independent Judiciary in an Interest-Group Perspective*, 18 JOURNAL OF LAW & ECONOMICS 875, 877 (1975); Merrill, *supra* note 52, at 1043; Seidenfeld, *supra* note 340.

377 See JERRY MASHAW, GREED, CHAOS, AND GOVERNANCE: USING PUBLIC CHOICE TO IMPROVE PUBLIC LAW 6-25 (Yale University Press 1997).

copyright holders in the decision-making process. As the competent authority in charge of broadcasting and television, the SARFT intended to maximize its departmental interest. It did so in a way that represented a parochial interest of a regulated community which was at odds with the public purposes reflected in the copyright statutory scheme. Thus, the former Article 43 was fiercely criticized as a method of subsidizing departmental interests at the expense of the numerous copyright holders.

The ostensibly unfair Article 43 existed for more than 10 years until it was ultimately modified in 2001. During that decade, the economic and social environment changed dramatically. The most distinct alteration unfolded in two aspects. On the one hand, the implementation of international treaties generated super national treatment on foreign copyright holders. Since joining the Berne Convention (1992) and the WCT (2007), China has applied international regulations which adopted statutory licenses for the broadcasting of foreign works,[378] whereas it still applied the former Article 43's fair use approach towards domestic works. The discriminating treatment generated an unfair situation where the protection level of foreign copyright holders was much higher than that of domestic ones, which went against domestic cultural creation and innovation.

On the other hand, the Chinese economic system shifted from a planned economy to a market economy. The radio and television stations were no longer absolute noncommercial public organizations and their abundant advertisement income could be sufficient to pay for copyright remuneration.

During the intense contest on retaining or abolishing Article 43, the legislator – Education, Science, Culture and Public Health Committee of the NPC – explicitly indicated that departmental interests should be subserviant to the national interest in order to foster a good environment for IPRs protection in society as a whole, and supported the NCAC modifying Article 43.[379] Notwithstanding this victory in the battle on legislation, the remuneration to the copyright holders by the radio and television stations remained a fictitious bill for 8 years,[380] until a compromise between the NCAC and the SARFT was finally reached in 2009.[381]

378 See Article 11 *bis* (2) of the Berne Convention (providing that compulsory licenses of broadcasting rights should not be prejudicial to the copyright holder's right to obtain equitable remuneration).

379 See Zheng Chengsi, Woguo zhuzuoquan fa xiuzheng qianyi [*Comment on the Revision of the Chinese Copyright Law*], 6 Zhuzuoquan [COPYRIGHT] 35, 42 (2001).

380 Article 43 of the 2001 Chinese Copyright Law provides: "A radio station or television station that broadcasts a published sound recording does not need to obtain permission from, but shall pay remuneration to the copyright owner, unless the parties concerned

Arguably, the phenomenon of departmental interest has been noticed and criticized in both the theoretical and practical fields for a long time, but there is still a long way to go before it can be solved systematically. Several complicated factors constitute the underpinning of departmental interest represented in the legislative process. First, the administrative agencies are actually undertaking the drafting of the law. Although the constitution and other laws stipulate wide sources of proposals, most proposals of law and regulations are provided by government agencies. Administrative agencies dominate the link in drafting and proposing which to a great extent influences the legislative process, and inevitably induces the departmental interest to filter into the legislation.

Second, despite having no authority to consider or vote on the draft, in fact administrative agencies can still exert an important influence on the legislature's behavior. On the one hand, the political status of administrative agencies in the state organ system compels the legislature to sufficiently emphasize and consider their opinions. In addition, the legislature "typically lacks specialized information and expertise," [382] thus is easily influenced by administrative "expertise". On the other hand, given that the implementation task of the legislation will ultimately depend on interrelated administrative departments, if a certain draft is opposed by the administrative agencies, the legislature would inevitably worry about its implementation after being enacted.

Third, given the assumption that the high per capita minority often prevails over the dormant low per capita majority in the legislative process, [383] the participation of the public in the legislative process customarily fails to compete with the administrative departments. Many approaches have been adopted by the NPC and its Standing Committee to hear the voices besides the departments',

---

have agreed otherwise. The specific measures shall be provided by the State Council." The State Council had not, however, promulgated any corresponding measures until 2009. In practice, corresponding stations refuse to pay any remuneration to copyright holders. Professor Zhou Lin criticizes the behavior as copyright infringement. See Zhou Lin, Zhuzuoquan bu jinjin shi "zuozhe zhi quan" [*Copyright Is not Merely the "Author's Right"*], Zhongguo faxue wang [China Legal Science Website], available at http://www.iolaw.org.cn/showNews.asp?id=643 (last visited Feb. 13, 2010).

381 Guangbo diantai dianshitai bofang luyin zhipin zhifu baochou zanxing banfa [Provisional Measures on the Copyright Remuneration Payment of Broadcasting Sound Recordings by Radio Stations and Television Stations] was enacted on Nov. 10, 2009 and implemented from Jan. 1, 2010. Subsequently, the directors of the Legislative Affairs Office of the State Council, the NCAC and the SARFT together attended the press conference to answer questions on the Provisional Measures, available at http://www.gov.cn/zwhd/2009-11/17/content_1466745.htm (last visited Feb. 11, 2010).

382 See Merrill, *supra* note 52, at 1049.

383 See KOMESAR, IMPERFECT ALTERNATIVES, *supra* note 51.

especially to solicit opinions from the ground for some important drafts. However, the influence of social forces and their counterparts can hardly contend with the powerful departments to avoid the unfair phenomenon where dominant agencies expand departmental power and reduce departmental liability, while aggravating the obligations and weakening the rights of counterparts in the legislative process.[384] Thus, the existence of the aforementioned factors increases the difficulty of the legislature in overcoming departmental protection when constituting a law.

2.2.2.2.3 Institutional Disjunction: the Dislocation between Administrative Protection and Judicial Protection

The original purpose of the "dual system" was to overcome the disadvantages of a single approach and to obtain mutual complementation. However, in practice the disjunction between administration and judiciary impairs the efficiency of the "dual system" and induces the phenomenon of "four excesses and four lacks": there are more IPR infringement cases than there are investigations into such cases; among the investigated cases, those receiving administrative punishments are greater than those handed over to the court and charged with criminal liability; more accessories are being punished than principals; and more light punishments such as probation are being enforced than are heavier punishments.[385]

Arguably, establishing and improving the junction mechanism between the judicial process and administrative protection are an inherent requirement for the development of a socialist market economy, maintaining the unification of the socialist legal system, and achieving social justice. As early as April 2001, the State Council definitely indicated in the Decision of the State Council on Rectifying and standardizing the Order in the Market Economy that the junction between administrative protection and criminal justice should be strengthened; cooperation mechanisms of information sharing, communication intercourse and early-warning against commercial crimes should be established.

In July 2001, the State Council enacted the Regulations on Referral of Suspected Criminal Cases by Administrative Enforcement Authorities (Order No.310 of the State Council). The regulation establishes the basic framework of a junction system, including the legal basis, procedure, duration, jurisdiction and legal liability of case referral. It explicitly stipulates that for those cases which administrative agencies should refer to public security organs, administrative punishments can not replace referral; administrative agencies which refer

384 See Liu, *supra* note 371, at 40-41.

385 See Cai Junyong, *supra* note 331.

criminal suspected cases should receive supervision under the procuratorial organ and supervisory organ.

Moreover, various administrative agencies have also created provisional regulations concerning the junction between administrative protection and criminal justice in special fields. For instance, in the copyright protection realm, the Ministry of Public Security and the NCAC jointly created the Provisional Regulations on Strengthening Junctions and Cooperation to Crack down on Copyright Crimes.[386]

Notwithstanding such regulations, on account of the lack of efficient connections and information sharing platforms, the due process requirement and the necessary supervision of administrative behavior, considerable numbers of infringement cases which are supposed to be delivered to the court and charged with criminal liability are now being dealt with through the application of administrative penalties. Yet pecuniary penalties alone are not sufficient to fill a deterrence role that can deter potential lawbreaking and crime. Rather it can foster infringement behavior. In this sense, the disjunction within the "dual system", to a large extent, weakens the effectiveness of copyright protection.

2.2.3 Fostering a Culture of Copyright in Chinese Society: A Long Way to Go

As Harold Berman, a well-known US jurist, once said, law must be believed in, otherwise it will exist in name only.[387] A law-abiding society demands not only perfect black letter law, but also law-abiding citizens – the latter perhaps being much more important than the former. "Legal consciousness, a basic recognition of the importance of law and of the relationship of law and legality to the development of other aspects of society, must be nurtured along with the growth of the legal system."[388] However, fostering a culture of copyright in Chinese society is not an easy job on account of various reasons rooted in historical, political, cultural, and social aspects.

---

386 Available at http://laws.ipr.gov.cn/ipr/laws/info/Article.jsp?a_no=16383&col_no=97&dir=200609 (last visited Feb. 8, 2010).

387 See Harold J. Berman, Falü yu zongjiao [LAW AND RELIGION] 26 (Liang Zhiping trans., Beijing: Shangwu yinshuguan [The Commercial Press] 2002).

388 James V. Feinerman, *The Rule of Law Imposed from Outside: China's Foreign-Oriented Legal Regime Since 1978*, in THE LIMITS OF THE RULE OF LAW IN CHINA 318 (Karen G. Turner, James V. Feinerman, & R. Kent Guy eds., University of Washington Press 2000).

First of all, there is no counterpart in Chinese culture to the Western concept of IP.[389] The private rights concept of copyright is a new perception that has only begun to be accepted in recent decades. Historically, there was no private copyright notion in ancient China.[390] "[T]he Confucian disdain for the profit motive prevailed over the capitalistic drive which propelled intellectual property reform in the West."[391] Following the ancient Chinese culture, writers and artists created works for edification and moral renewal rather than for profit. As a famed Chinese aphorism suggests, "Genuine scholars let the later world discover their work [rather than promulgate and profit from it themselves]."[392]

Additionally, for a long time after the establishment of the PRC, the private copyright of the author, including the ownership and remuneration right, was regarded as departing from the socialist ideal of public ownership, and thus suffered suppression during various political movements against "capitalist privatization".[393]

After the 3rd plenary session of the 11th central committee of the CPC in 1978, Deng Xiaoping suggested that there was no conflict between the socialism and the market economy.[394] In order to develop the socialist economy, we need to combine the planned economy and the market economy together.[395] Deng Xiaoping further clarified the relationship between the two categories of economy in his famous talk during inspecting the south:

> Having more plan, or more market, is not the essential distinction between socialism and capitalism. The planned economy is not the equivalent of socialism, capitalism also has plan; and the market economy is not the equivalent of capitalism, socialism also has market. Both plan and market are economic instruments.[396]

---

389 See ALFORD, *supra* note 291, at 2.

390 See *supra* section 2.2.1.1.

391 Guy Yonay, *Book Note: To Steal A Book Is an Elegant Offense: Intellectual Property Law in Chinese Civilization*, 8 HARV. J. L. & TECH. 537, 538 (1995).

392 ALFORD, *supra* note 291, at 29.

393 The political movements against "capitalist privatization mainly include the "Hu Feng event" in 1955, the "Anti-rightist Movement" in 1957, the "Great Leap Forward Movement" from 1958-1960, the "Rectification Movement" in 1964, and especially the "Cultural Revolution" from 1966-1967. For more details about the zigzag evolution of copyright policy in the PRC, see *supra* section 2.2.1.2.

394 See Shehui zhuyi ye keyi gao shichang jingji [*Socialism also Has a Market*], in Deng Xiaoping wenxuan [SELECTED WORKS OF DENG XIAOPING] (vol. 2, Beijing: Renmin chubanshe [People's Publishing House] 1983).

395 See Shehui zhuyi he shichang jingji bu cunzai genben maodun [*No Conflict Exists Between Markets and Socialism*], in Deng Xiaoping wenxuan [SELECTED WORKS OF DENG XIAOPING] (vol. 3, Beijing: Ba yi chubanshe [Ba Yi Press] 1993).

396 Deng Xiaoping wenxuan [SELECTED WORKS OF DENG XIAOPING], *id.* at 373.

Consequently, copyright has been acknowledged as the author' private right. Through years of effort, China has established the legislative framework of copyright protection. However, the dangers of formalism and the gap between official pronouncement and the internalization of policy shifts still cannot be overlooked. "Indeed, denunciation of formalism in work style remains a constant feature of Chinese social commentary."[397]

Law is a necessity, but not in itself sufficient. In order to employ the law as a dynamic force rather than merely a set of abstract rules disconnected from the changing society surrounding it, an acknowledgement of the complex and often contradictory relationship between law and economic change is necessary. Moreover, a consideration of the connections between legislation and the practices it seeks to induce or to alter is essential, along with attention to the resistance that legal change often meets and the reasons therefore.[398]

If we employ a cloud-built framework fabricated by legislators to rule the masses, the upper design layer will become the more important custom of rules lying atop the infrastructure of practice. Consequently, formalism would be inevitably generated.[399] On account of the disjunction between the artificial national superstructure and the under layer of the practical customs of the citizens, this formalistic law exists in name only. As Oliver Wendell Holmes has indicated, the most crucial character of a sound law is its responses to the actual perceptions and demands of the people, no matter whether these perceptions or demands be correct or not.[400] Therefore, in order to prevent the law from decaying into mere logical symbols on paper, we must always pay attention to the operation of law rather than the ideal of law.[401]

Moreover, lawmaking is one thing; the forming of a copyright culture among citizens is another. Arguably, rampant copyright infringements are the result of institutional causes, rather than the imperfectness of the law. As a famous Chinese saying indicated, "to steal a book is an elegant offense". The reason for this comes from the deep legal consciousness perspective. In ancient China, piracy had seldom been conceived of as an illegal act akin to theft.[402]

---

397 Feinerman, *supra* note 388, at 318.

398 See *id.*

399 Huang Renyu [Ray Huang], Zhongguo da lishi [CHINA: A MACRO HISTORY] 15 (Beijing: San lian shudian [San Lian Bookstore] 1997).

400 See Su Li, Songfa xiaxiang [SENDING THE LAW TO THE COUNTRYSIDE] 285 (Beijing: Zhongguo zhengfa daxue chubanshe [China University of Political Science and Law Press] 2000).

401 See Su Li, Zhidu shi ruhe xingcheng de [HOW A SYSTEM IS FORMULATED] 158 (Guangzhou: Zhongshan daxue chubanshe [Sun Yat-sen University Press] 1999).

402 See Yi Yufeng, *supra* note 255, at 174.

Although now piracy is forbidden by copyright law, there still remains some gap between the regulation and people's reaction towards piracy.

As Ewick and Silbey indicate, there are three categories of attitudes that suggest how common people approach the law: before the law, with the law, and against the law.

> [T]hose who operate before the law, see the law as functioning in a way that provides an external order to everyday experiences, regulating ordinary interactions with institutions and authority figures. People who see themselves as functioning "before the law" abide by and cooperate with the law and figures that represent the law. Individuals working "with the law" use the law strategically in ways that beneficially suit their interests. People who work with the law often draw on previous legal experience – successful or not – to help inform their outlook or provide perspective on how to deal with a wide range of problems.... [T]hese individuals ... "playing" with the law and toying with legality in a way that reflects a game-like approach in which these individuals gauge wins and losses and strategic next steps, manipulating the law and taking pride in doing so when the outcomes are favorable to them. A final group of individuals were found to work "against the law." Those working against the law did not engage the law on a regular basis and if they did, they often encountered the law in a way that was off-putting and caused a significant amount of stress.... Moreover, among those categorized as operating "against the law" there is an important subgroup that avoids the law almost entirely and often deliberately. Those who avoid the law can be described as expressing a particular resignation to engaging with the law due to perceptions of accessibility, fairness, or relevance to the everyday experience they encounter.[403]

Therefore, in order to rectify the attitude against the law, the abiding of copyright law should focus on cultivating citizen's legal consciousness. The establishment of the legal consciousness of copyright protection among citizens is a crucial but never easy task. Looking back over the years since the founding of the PRC, several generations of people have been lacking in this respect. It is now high time to educate our citizens to be well qualified in the principles of IPRs protection. Notably, fostering a culture of IPRs has been emphasized by the Outline of the National IP Strategy as one of its strategic focuses:

> Strengthen the knowledge propagation on intellectual property right and increase the awareness of intellectual property right in the whole society. Carry out the ordinary intellectual property right education extensively. Increase the intellectual property right content in the national promotion of the public moral culture and the national ordinary education in law. By advocating the moral concepts of being proud of

---

403 Diana Hernández, *"I'M Gonna Call My Lawyer:" Shifting Legal Consciousness at the Intersection of Inequality*, in SPECIAL ISSUE: INTERDISCIPLINARY LEGAL STUDIES: THE NEXT GENERATION 98-99 (Austin Sarat ed., UK: Emerald Group Publishing Limited 2010).

> innovation and honesty and ashamed of plagiarism and counterfeiting or cheating, the intellectual property right culture could be established throughout China, characterized by respect for knowledge, enthusiasm for innovation and creation, being honest and obeying law.[404]

More specifically, the Outline of the National IP Strategy provides measures for promoting the cultivation of an intellectual property culture:

> [First,] Set up a working mechanism for publicizing information about intellectual property that is led by the government and supported by the media, in which the public widely participates. The coordination mechanism needs to be improved and relevant policies and working plans need to be formulated to promote public awareness of intellectual property and the development of an intellectual property culture.[405]
>
> [Second,] Offer intellectual property courses in higher education institutions and to introduce education on intellectual property into the quality-based education to students of such institutions. A general education plan on intellectual property needs to be formulated and implemented, and intellectual property education needs to be listed into teaching plans in primary and high schools in whole China.[406]

In a word, there is a long way to go before the effective fulfillment of copyright protection in Chinese society becomes a reality.[407] As Roscoe Pound has indicated the term "law" has many meanings, and any system of law is much more than simply a body of laws. "Law shows us how to fill in the gaps of laws."[408] "Only when this understanding of the role of law is more widely understood in the PRC will the threat posed by legal formalism to social and economic development be removed."[409] And only when everyone self-consciously abides copyright law will piracy and other infringements die out.

2.2.4 Pricing: A Vital Factor Influencing the Buying Behavior of Piracy Products in the Chinese Market

In addition to several reasons such as the law making process, the implementation of the law, and adherence to the law causing the piracy dilemma

---

404 Article 15 of the Outline of the National IP Strategy, available at http://www.gov.cn/english/2008-06/21/content_1023471.htm (last visited Mar. 11, 2010).

405 *Id.* Article 63.

406 *Id.* Article 64.

407 See Song Muwen, *supra* note 292, at 4.

408 Roscoe Pound, *Development of Law in Modern China*, 2 SINO-AMERICAN RELATIONS (Taipei) 60, 60 (1976).

409 Feinerman, *supra* note 388, at 319.

in China, from the market perspective, pricing remains a vital factor influencing the buying behavior of pirated products.

Arguably, the economic source of piracy is the market demand and market supply foundations. On the one hand, consumers prefer to choose pirated copies because the consumption value of pirated copies is higher than legal ones. When making the decision to purchase, consumers do not consider the price or quality separately, but rather the cost and performance ratio. Although there is some gap regarding the quality and service between the pirate and the legal edition, the difference is much smaller than in the price. In the condition of the same consumption budget, the marginal utility curve of buying a pirated version is higher than buying a legal product.[410]

On the other hand, producers prefer to produce pirated copies because the productive value of pirates is greater than legal editions. The profit of manufacturers depends on the balance between the product price and production cost. By virtue of authorizing and protecting private rights, IP laws provide incentives to promote further creativity. At the same time, it creates a monopoly on producing legal copies and thus raises the cost of those who want to produce legal editions. In contrast, the production of pirates is cheaper not only due to the lack of R&D costs, but also because producers usually use cheaper materials and simpler working procedures.[411]

Generally speaking, in light of the average income level of the Chinese population, the pricing of legal editions is too high to be affordable to the average person. Although as a whole China has overtaken Japan as world's second-biggest economy, the average Chinese person is much poorer than the average Japanese person. According to an estimate of the International Monetary Fund (IMF), the GDP per head of the population is around $34,000 in Japan, while it is only $7,500 in China.[412]

"Most people in China are still poor, more people live in the countryside than in cities."[413] For instance, the basic living allowance for citizens in China is around 200 RMB (around $30).[414] Take the pricing of a popular movie – *Inception* for example. The legal copy of a DVD is about 50 RMB, almost ¼ of the allowance. It would not be an exaggeration to say that legal edition is a

410 See Wei Tao, Daoban xianxiang de jingji genyuan tanxi [*Research on the Economic Root of Piracy*], 14 Jinrong jingji [FINANCE & ECONOMY] 120, 120 (2009).

411 See *id.* at 120-21.

412 See *China Overtakes Japan as World's Second-biggest Economy*, BBC NEWS, http://www.bbc.co.uk/news/business-12427321 (last visited Feb. 16, 2011).

413 See *id.*

414 The RMB/US dollar exchange rate was around 6.6 in Feb. 2011.

luxury for those people living depending on relief. In comparison, the pirated copy is less than 5 RMB which seems more reasonable and affordable.

Similarly, the Microsoft Windows Server 2003 was priced at 11000 RMB, and Microsoft Office 2003 priced at 3500 RMB. In contrast, the pirated edition was less than 10 RMB.[415] Is it reasonable if it costs a Chinese person a whole year's salary to buy such legal software? Although Microsoft has reduced the price in recent years, the pirated edition still retains an overwhelming advantage on price. For instance, in 2007 Microsoft Office sold for 1500 RMB and Windows XP sold for 1990 RMB, while the pirated edition of each only sold for 10 RMB or even less. Thus, it is no wonder that consumers would choose the latter.[416] To take another example, the pirated "Tomato Windows XP" was even more popular on account of being freely downloadable, canceling the validation procedure of Microsoft, and uninstalling unusual functions of the original Windows XP.[417]

Therefore, the key issue of the piracy phenomenon is the price. Without solving this problem, any anti-piracy measures would only treat the symptoms rather than the source.[418] By virtue of reducing the price of legal editions to make them obtain a better cost performance ratio, legal products will ultimately replace pirates in the market.

---

415 See Jiage qishi poshi women shiyong daoban [*Price Discrimination Forces Us to Use Piracy*], available at http://tech.163.com/special/000915RB/daoban.html (last visited Feb. 12, 2010).

416 See Tan Ying, Ruanjian daoban chengyin yu zhili jizhi fenxi [*Analysis of the Reasons for Software Piracy and Governance Mechanisms*], 1 Ruanjian daokan [SOFTWARE GUIDE] 3, 3 (2009).

417 According to a survey including 7037 netizens, almost 88.4% stated they had used "Tomato Windows XP" or similar pirated operation systems. More than 100 million computers had installed "Tomato Windows XP". Before being banned "Tomato Windows XP" occupied the largest share of market. See Weiruan "shouwang" daoban Windows fanqie huayuan zhanzhang bei ju [*Microsoft Beats Piracy: Directors of the Tomato Windows XP Were Arrested*], available at http://news.xinhuanet.com/tech/2008-08/21/content_9564180.htm; in 2009, "Tomato Windows XP" was banned as pirate, and directors of the "Tomato" website were sentenced criminal liabilities. See Zhongguo daji daoban "fanqie huayuan" an ruxuan shida anli [*China Beats Piracy: "Tomato" Case Is Selected as One of the Ten Classic Cases*], available at http://news.xinhuanet.com/tech/2010-04/23/content_13410059.htm (last visited Feb. 11, 2011).

418 See Shi Mingxia, Ruanjian daoban yu fan daoban de jiage sikao [*Rethinking Price in Software Piracy and Anti-piracy*], available at http://www.jgsc.gov.cn/2005-6/200562164747.htm (last visited Feb. 6, 2011).

In order to do so, first of all, copyright holders should set a reasonable price. The current pricing which is similar to the sale of fashionable dresses does not work well in China. This type of pricing follows a model wherein very high prices are charged upon entering the market in order to recover the initial investment as soon as possible, then prices are gradually lowered, ultimately ending with a liquidation sale. However, this kind of pricing strategy scares away a large number of potential customers at the very start. They might choose the much cheaper pirated editions instead. Even later a cheaper legal version will lose its attraction on account of almost all potential clients having already purchased pirated editions.[419] Consequently, it enters into a vicious circle that any legal products could not be sold in large quantities on account of the ubiquitous pirates. And usually, the better the products, the more the pirates. The insufficient sale of products impels copyright holders to sell them at a high price to recover their costs. However, on the other hand, the high price obstructs the sale of legal editions and boosts the pirates.[420]

There are several approaches to overcome this vicious circle. First, right holders should change their biased view that "there is no cheap legal edition, and cheap editions must be illegal". For a long time, copyright holders have treated legal editions as top quality articles and artificially separated them from pirates on price. Instead, right holders should set up an affordable price according to the supply and demand in the market. Second, right holders should reduce costs to the utmost, including those costs in packaging, advertising, and transportation. Only through diminishing the cost and sale price, and shortening the price difference between the legal and illegal ones, could legal editions seize the market and depress the probability of piracy. Finally, right holders could increase the number of first sales both to reduce the cost and to occupy the market.[421]

Fortunately, more and more copyright holders have realized that pricing is the vital factor of piracy and started to positively reduce their prices rather than merely waiting for the government's anti-piracy as before. A famous example of "selling legal product at a pirated price" was when the Richpeace Group reduced its "embroidery software" from 30000 RMB to 300 RMB. It was a self-help relief after years of unsuccessful legal relief. It was not only costly and time-consuming to sue pirates, but they also had to deal with the failure to receive compensation even after winning lawsuits. The fact was that either the pirate ran

419 See *id.*

420 See Liu Xingcheng, Zhongguo fan daoban de pingjing [*The Bottleneck of Anti-piracy in China*], 7 Faren [FAREN MAGAZINE] 43, 44 (2009).

421 See *id.*

away or was too poor to have any property to recover against. In an ironical case, the infringer was unable to afford the compensation and thus sincerely requested to work in the Richpeace Group and to pay the damager in labor. Later, the Richpeace Group decided to occupy the market by virtue of reducing prices. This strategy works well in practice. More and more customers not only choose to buy the legal edition, but also begin to trust the company and buy related equipment from them.[422]

In addition, in recent years, the price of the audio and video products has reduced sharply. For instance, in 2009 the price of a legal DVD in Lanzhou, Gan Su province had decreased 60% compared to the year before. The quality of legal editions of HDVD was much higher than the pirates, while the price difference was only a little. Thus, the low-cost legal editions became very popular among citizens. A customer mentioned that, in the past it was too expensive to buy legal DVDs. Since the price difference was now much smaller, most people began to buy legal copies instead.[423] Although at present the price difference still remains to some extent, if the market can maintain this positive cycle, there will be more space for price decreases of legal products.

Moreover, as the development of the national economy and the increase of residents' income progress, more and more people in the high and medium income groups will prefer to consume legal products. It is estimated that more than 60% of Chinese citizens have shown their preference for legal editions.[424] For example, the movie *Inception* has received more than 450 million RMB at the box office in the Chinese market.[425] We can imagine how many people chose the legal edition of this film over the pirate.

In a word, it is not that the Chinese people inherently like pirates, but that they have chosen to purchase them because of their income and the price. "Only when the piracy bears high cost, low profit, and great risk, while the anti-piracy

---

422 See Ma Li, "Daoban jia" beihou de xuanji [*Secret behind the Price of Piracy*], 8 Faren [FAREN MAGAZINE] 52, 52-53 (2009).

423 See Ge Qiang, Gansu Lanzhoushi zhengban guangdie jiage tongbi xiajiang 60% [*In Lanzhou, Gan Su province the Price of Legal DVD has decreased 60% comparing to the year before*], available at http://news.ipr.gov.cn/ipr/news/info/Article.jsp?a_no=289726&col_no=2&dir=200908 (last visited Feb. 13, 2011).

424 See Liu Yang, Fenxi jigou: jiangdi ruanjian jiage caineng cong genben shang jiejue daoban [*Analysis Institute: Only Reducing the Software Price Can Fundamentally Solve Piracy*], available at http://tech.163.com/06/0524/08/2HSHMH9L000915BD.html (last visited Feb. 13, 2011).

425 See Yu Deshu, "Dao meng kongjian" zaidu yanqi xiaxian guonei zong piaofang chao 4.5 yi [*Movie Inception has received more than 450 million RMB for box office in the Chinese market*], available at http://ent.qq.com/a/20101105/000759.htm (last visited Feb. 15, 2011).

receives low cost, high profit, and small risk, the bottleneck of anti-piracy could be overcome, and piracy will be no longer a problem."[426]

2.3 Summary

Arguably, the cause of abandoning the "free zone" of private use in the digital copyright framework is the outcome of anti-piracy. When creating the 2006 Transmission Regulation, Chinese legislators were faced with two-tier pressures: the pressure against rampant piracy, and the pressure of keeping in line with international IPRs conventions. As large-scale P2P file sharing by end users greatly undermined copyright incentives, and users sought protection via the private use exemption, legislators adopted an oversimplified and crude way to exclude private use from the scope of fair use of digital copyright.

However, the abolishment of the private use limitation in the digital copyright framework as an anti-piracy measure has failed to catch the deep-seated causes of piracy dilemma in China. The actual underpinning factors are to be found in the law making process, the implementation of the law, adherence to the law, and also in the market. First, the copyright system is an exotic thing in China. From the first neoteric Chinese copyright law created at gunpoint, to the ups and downs of copyright policy in the PRC, as well as the difficult foundation of a socialist copyright law, the twists and turns remind us that the effective implementation and self-conscious abiding of the copyright law could not be accomplished in one action.

Second, administrative protection currently dominates the copyright implementation "dual system", and induces multiple problems, such as agency overlap, minoritarian bias, and institutional disjunction. The limitations, to some extent, derogate from the effect of copyright protection and anti-piracy efforts.

Third, on account of the fact that there is no counterpart in Chinese culture to the Western concept of IP, and that the private copyright of the author was not recognized for a long time after the establishment of the PRC, there is a long way to go before fostering a culture of copyright in Chinese society, and cultivating citizen's legal consciousness to self-consciously abide by the law.

Last but not least, from the market perspective, the pricing of legal editions is too high for the average Chinese citizen to afford. Only by virtue of reducing the price of legal editions to give them a better cost performance ratio, and increasing resident's income alongside the development of the national economy, will Chinese people choose, and be able to afford legal products.

426 Liu Xingcheng, *supra* note 420, at 43.

Chapter III Pros and Cons: Debates on Employing TPMs to Control Private Use in the Digital Environment

Digital technology and Internet networks have led to a "digital dilemma". On the one hand, they have greatly facilitated the transmission of information and enriched end users' enjoyment. On the other hand, however, they have diminished the control of right holders over their copyrighted works.[427]

Given the fact that the abolishment of the private use limitation in the Chinese digital copyright framework as an anti-piracy measure is not a sound choice, in this chapter, I intend to explore the right place for private use in the digital copyright regime. I will start with the boundary between copyright exploitation and private use enjoyment. Then I will discuss the efforts to harmonize TPMs and private use from both the legislative and judicial perspectives, as well as the tit-for-tat debates on employing TPMs to control private use. Finally, I will employ the objective of the Chinese copyright law to evaluate the TPMs control and private use.

3.1 Locating the Boundary between Exploitation and Enjoyment

As mentioned above, private use is the use of works by individual users for their own study, research, or self-entertainment purposes, such as reading a book, singing a song, and watching a movie on television. On account of its being limited to the enjoyment of works and not their exploitation, private use has traditionally been outside of the scope of copyright.

Private use is, however, notoriously uncertain both in theory and in application. Given the fact that the copying itself, be it in a private use context or a piracy context, is physically the same; it is very easy to confuse private copying with piracy. And it is a fact that a particular private copying might itself

427 See COMMITTEE ON INTELLECTUAL PROPERTY RIGHTS AND THE EMERGING INFORMATION INFRASTRUCTURE ET AL., THE DIGITAL DILEMMA: INTELLECTUAL PROPERTY IN THE INFORMATION AGE (US: National Academies Press 2000); Raymond Shih Ray Ku, *The Creative Destruction of Copyright: Napster and the New Economics of Digital Technology*, 69 U. CHI. L. REV. 263, 270-76 (2002).

substitute for the purchase of a copy and thus diminish the income which copyright holders would otherwise receive.

But on the other hand, they are different in essence. Private copies are used for non-commercial private purposes and limited to the private sphere, while pirated copies enter into public distribution and are normally made for purposes of commercial exploitation. For example, "[a]n individual who rips a CD to her iPod,[428] turns on her brother's computer, fast forwards through objectionable portions of a television show or DVD, or plays music with the windows open isn't violating the copyright law".[429] Copyright should to be "interpreted subject to the understanding that copyright prohibits unauthorized exploitation but not unauthorized enjoyment".[430] The difficulty, however, is how to separate them in practice.

There are at least two vital sub-factors which need to be taken into consideration when locating the boundary between exploitation and enjoyment. These are whether the activity involves commercial or noncommercial use, and whether it falls within the market place or the private domain. In this section, I intend to analyze these sub-factors as well as the difficulty in telling the difference between them in reality.

3.1.1 Commercial Use or Noncommercial Use

In respect of private use, the nature of the use must be non-commercial. Maryly Snow indicates that "'personal use' should be considered an antonym of 'commercial use'."[431] For example, the first factor of the fair use test in Section 107 of the US Copyright Act emphasizes the character of the use, including whether such use is of a commercial nature or is for non-profit educational

428 See Fred von Lohmann, *RIAA Says Ripping CDs to Your iPod Is Not Fair Use*, Deep Links, Feb. 15, 2006, available at http://www.eff.org/deeplinks/archives/00409.php (comparing that in the Grosker case, RIAA stated that ripping a CD was "perfectly lawful") (last visited Oct. 6, 2010).

429 Litman, *supra* note 4, at 1908.

430 *Id.*

431 Rober A. Baron, *Editor's Introduction, Summary and Analysis*, in COPYRIGHT AND FAIR USE: THE GREAT IMAGE DEBATE 3-4, 255 (Vol. 12, Robert A. Barton ed., Netherlands: Gordon & Breach Publishers 1997).

purposes.[432] A finding of commercial use weighs against, although it does not preclude a determination of fairness.[433]

3.1.1.1 Commercial Use Effecting on Exploitation

Generally speaking, a commercial use is a use which is undertaken for a business purpose, rather than for recreational, hobby, educational, or other purposes. Such uses are usually attributed to a for-profit entity. Thus, uses by for-profit companies are typically considered more commercial, while uses by an individual, university or other educational institution, or non-profit organizations, such as public libraries, charities, and other organizations aiming at promoting public social welfare, are less commercial although not decidedly noncommercial.[434]

Commercial uses have traditionally been denied fair use privileges.[435] The reason to forbid unauthorized commercial use is that it harms copyright holders' exploitation. For example, in *Sony Corp. v. Universal City Studios, Inc.*, the Court unequivocally held that, "every commercial use of copyrighted material is presumptively an unfair exploitation of the monopoly privilege that belongs to the owner of the copyright".[436]

The economic rights granted to right holders ensure them the ability to exploit their works in the market place and to exclude others from the exploitation of their works.[437] By doing so, copyright law safeguards the ability of right holders to reap an economic reward which is sufficient to cover the costs of producing the work, and to promote incentives for further creation.

---

432 See 17 U.S.C. § 107.

433 See *Campbell, Aka Skyywalker, et al. v. Acuff-Rose Music Inc.*, 510 U.S. 569, 584 (1993).

434 See Creative Commons, *Defining "Noncommercial": A Study of How the Online Population Understands "Noncommercial Use"*, Sep. 2009, p.11, available at http://mirrors.creativecommons.org/defining-noncommercial/Defining_Noncommercial_fullreport.pdf (last visited Oct. 6, 2010).

435 See WILLIAM F. PATRY, THE FAIR USE PRIVILEGE IN COPYRIGHT LAW 365 (Washington, D.C.: The Bureau of National Affairs, Inc. 1985).

436 *Sony Corp. of America v. Universal City Studio Inc.* 104 S. Ct. 774, 793 (1984).

437 For more details about the economic rights that the copyright law confers on the copyright holder, see *supra* section 1.1.2.1.

3.1.1.2 Noncommercial Use for Enjoyment

Strictly speaking, to provide a concrete definition of "non-commercial" is very difficult.[438] A most familiar example might be the Non-Commercial use (NC) clause of the Creative Commons (CC). Yet it looks a little bit too simple: "You may not use the work for commercial purposes."[439] Certainly the full legal code behind the CC Non-Commercial attribute is more concrete: "You may not exercise any of the rights granted to You... above in any manner that is primarily intended for or directed toward commercial advantage or private monetary compensation."[440] This is a negative definition to limit the scope of licensee's rights, but we still can not be sure what non-commercial is supposed to mean.[441]

One of the important reasons that is in support of an exemption of private use of copyrighted works from the realm of copyright control is, "copyright's main job is to provide authors with a means of controlling commercial exploitations of their works, as these are the acts by which copyright owners typically recoup their investments in producing their creative works."[442] Private uses, however, do not involve such exploitations but rather relate to enjoyment.

3.1.1.3 A Plausible Boundary between Commercial Use and Noncommercial Use

In spite of the aforementioned definitions, to determine whether a certain use is commercial or noncommercial still remains a difficult task in practice. For example, in 2009 a report issued by Creative Commons trying to define "noncommercial" through an empirical study of how the online population understands "noncommercial use" suggests that there is more uncertainty than clarity around whether specific uses are commercial or noncommercial.[443] A

438 See Adnan Hadzi, *Involve me, and I Will Understand: Introducing the Data Sphere*, in DEPTFORD. TV DIARIES VOLUME II: PIRATE STRATEGIES 164 (UK: Mute Publishing Ltd 2008).

439 See CC Attribution-NonCommercial 3.0 Unported, available at http://creativecommons.org/licenses/by-nc/3.0/ (last visited Oct. 6, 2010).

440 See section 4 (b) of *Creative Commons Legal Code*, available at http://creativecommons.org/licenses/by-nc-nd/2.5/legalcode (last visited Oct. 6, 2010).

441 See Mikael Pawlo, *What Is the Meaning of Non-Commercial?* available at http://fr.creativecommons.org/articles/sweden.htm#8 (last visited Oct. 8, 2010).

442 Pamela Samuelson et al., *supra* note 171, at 53.

443 See Creative Commons, *supra* note 434 (stating "uses that earn users money or involve online advertising to be commercial, while uses by organizations, by individuals, or for charitable purposes are less commercial but not decidedly noncommercial. Similarly,

similar observation was made by the House report when constituting the 1976 US Copyright Act:

> The line between commercial and "nonprofit" organizations is increasingly difficult to draw. Many "non-profit" organizations are highly subsidized and capable of paying royalties, and the widespread public exploitation of copyrighted works by public broadcasters and other noncommercial organizations is likely to grow. In addition to these trends, it is worth noting that performances and displays are continuing to supplant markets for printed copies and that in the future a broad "not for profit" exemption could not only hurt authors but could dry up their incentive to write.[444]

In addition, "[c]ourts repeatedly bog themselves down in determining whether a use is commercial or noncommercial. Finding a meaningful and consistent definition of commercial use has proved an elusive goal."[445] For example, in *A & M Records v. Napster*,[446] the Ninth Circuit held that the P2P file-sharing activities constituted commercial use because users received copyrighted material free which they would otherwise pay for. In other words, "repeated and exploitative unauthorized copies of copyrighted works were made to save the expense of purchasing authorized copies".[447]

In another case, the Ninth Circuit held that distributing 30,000 free copies of a piece of religious literature constituted a commercial use because the distributor profited from the use of works by attracting new members who would ultimately contribute to the organization by ways of tithing or donation.[448]

Following this logic, "all use is commercial use because, at some level, any unpaid use of a copyrighted work causes someone to lose potential revenue, and any use has some conceivable commercial motivation."[449] However, as Jessica Litman points out, if any use that makes a person get something for free which

---

uses by for-profit companies are typically considered more commercial.... there is more uncertainty than clarity around whether specific uses of online content are commercial or noncommercial."), available at http://mirrors.creativecommons.org/defining-noncommercial/Defining_Noncommercial_fullreport.pdf (last visited Oct. 10, 2010).

444 H. R. Rep. No. 94-1476, at 62-63.

445 John Tehranian, *Et Tu, Fair Use? The Triumph of Natural-Law Copyright*, 38 U.C. DAVIS L. REV. 465, 502 (2005).

446 *A & M Records, Inc. v. Napster, Inc.*, 239 F. 3d 1004 (9th Cir. 2001).

447 *Id.* at 1015.

448 See *Worldwide Church of God v. Philadelphia Church of God, Inc.*, 227 F. 3d 1110, 1117-18 (9th Cir. 2000), cert. denied, 532 U.S. 958 (2001).

449 Tehranian, *supra* note 445, at 502.

they would otherwise need to pay for is a commercial use, then most lawful unlicensed uses would be commercial.[450]

It is the right holders' insidious trick to replace the concept of "commercial use" with "unlicensed use", and "noncommercial use" with "licensed use". Accepting this would weaken the lawful copyright limitations and expand the limited exclusive copyright into an unlimited monopoly right. "Defining commercial use so broadly makes it useless as a sorting tool. In order to help us distinguish permissible from impermissible uses, we need to define commercial use narrowly enough to capture direct financial gain and exclude more indirect benefit."[451]

However, not all people would agree with the exclusion of indirect benefit. In *Pacific & Southern Co. v. Duncan*, the Eleventh Circuit held that, "[c]opyrights protect owners who immediately market a work no more stringently than owners who delay before entering the market". Thus, although the copyright holder did not presently exploit the same market developed by the defendant, the actual harmful effect denied the fair use defense.[452] Because the "actual harmful effect" referred not only to the loss of actual sales, but also the loss of sales that right holders would have exploited, but chose not to or delayed to.[453]

3.1.2 Market Place or Private Domain

When determining whether a certain use is private use, there is another sub-factor – the use must be within the private domain. In other words, the use of copyrighted works should be limited to the private sphere, and not be for public distribution or sale to others. This character of private domain draws a distinct line between private use and piracy.

3.1.2.1 Copyright as a Marketing Monopoly

Piracy is forbidden by copyright law and regarded as a serious copyright infringement which wipes out the profits of copyright holders and harms their incentives to create. Generally speaking, piracy describes the unauthorized

450 Litman, *supra* note 4, at 1913-14.

451 *Id.* at 1914.

452 *Pacific & Southern Co. v. Duncan*, 744 F. 2d 1490, 1496-97 (11th Cir. 1984).

453 See PATRY, *supra* note 435, at 368.

public distribution of copies of a copyrighted work.[454] From an economic perspective, a pirated work which sells at a relatively low price constitutes a competing substitute for a copyrighted work in the marketplace.

The reason why piracy is a copyright infringement is because it violates the right holder's exclusive right of public distribution. The right of public distribution denotes the right to distribute copies of copyrighted works to the public by sale or other transfer of ownership, or by rental, lease, or lending.[455] In other words, it means to put copyrighted material into public circulation. Notably, the distribution right is given "in respect of the issuing of each and every copy (including the original)", which needs to be distinguished from the publication right of making the works available to the public for the first time.[456]

3.1.2.2 Non-competitive Private Use in the Private Domain

In contrast, private use is limited to within the private domain. Thus, whether or not there is public distribution is a key element to separate private use from piracy.

A following related question is who constitutes the public? In general, public indicates more than a selected group of people.[457] It should be answered "in the same way as it has been in the cases concerning copyright infringement by performing the work in public".[458] To take an example, the US Copyright Act defines "public performance" as "to perform … it at a place open to the public or at any place where a substantial number of persons outside of a normal circle of a family and its social acquaintances is gathered".[459]

Similarly, in accordance with Section 15 Para. 2 of the German Copyright Act, "the author has the exclusive right to perform his work publicly in intangible form. Non-public performances (e.g. a private listening to a CD) are

454 See Michael J. Meurer, *Price Discrimination, Personal Use and Piracy: Copyright Protection of Digital Works*, 45 BUFF. L. REV. 845, 852 (1997) (The author only refers to piracy of literal copies in this article. However, the author also admits that unauthorized public distribution of some derivative works, e.g., translations of a factual work or textbook or unauthorized anthologies might reasonably be treated as piracy.).

455 See 17 U.S.C. § 106(3).

456 See BENTLY & SHERMAN, *supra* note 14, at 137 (Citing CDPA 1988 § 18(4) in respect of the distribution right in UK law). *Cf*. CDPA 1988 § 175 in respect of the publication right.

457 KENT D. STUCKEY, INTERNET AND ONLINE LAW 6-13 (Release 14, NY: Law Journal Press 2003).

458 HECTOR MACQUEEN ET AL, CONTEMPORARY INTELLECTUAL PROPERTY: LAW AND POLICY 140 (NY: Oxford University Press 2008).

459 17 U.S.C. § 101.

not subject to the author's exploitation rights. Performances are public if they are intended for a multitude of people which are not connected by personal relationship between each other or with the reproducer."[460]

In the case law on determining the character of the public, the selected group of people, such as the residents of a particular location, guests at a dinner party, and classmates at a university, are bound together by a link that distinguishes them from the general public.[461] For example, in *Duck v. Bates*, the Court of Appeal held that an amateur dramatic club which performed the play *Our Boys* at Guy's Hospital for the entertainment of the nurses did not infringe copyright.[462] This was because it was held to be a domestic performance, not a performance at a place open to the public, such as a shopping mall, restaurant or park, where anyone could enter and join.

Therefore, piracy refers to the distribution of copyrighted copies to more than a selected group of people. In contrast, the scope of private use is even narrower than the selected group of people, such as classmates, colleagues, wedding guests, or other ordinary social acquaintances, but denotes to only family members and close friends. In this sense, private use occurs in a sphere which is out of the control of the copyright holders' exclusive rights.

Additionally, when determining the private use, the sub-factors of both noncommercial and private domain should be satisfied. For example, in *A & M Records v. Napster*, [463] whether the court was correct in defining the "commercial" use might be questioned.[464] But the P2P file-sharing by Napster users constituted copyright infringement because such uses had already exceeded the private realm, and referred to the public distribution.

3.1.3 The Blurred Distinction between Enjoyment and Exploitation in the Digital Environment

The conventional boundary between enjoyment and exploitation has become more and more difficult to distinguish in the digital environment. New digital technology endows users with new utilities for copyrighted works, and the chance for copyright holders to profit from these utilities at the same time. In other words, it provides new abilities both for copyright holders to explore the new market and for users to enjoy the new utility.

---

460 WENDLER ET AL. EDS., *supra* note 15, at 191.

461 See BENTLY & SHERMAN, *supra* note 14, at 141-42.

462 See *Duck v. Bates*, (1884) 13 QBD 843.

463 *A & M Records, Inc. v. Napster, Inc.*, 239 F. 3d 1004 (9th Cir. 2001).

464 See *id.* at 1015.

For instance, one of the significant challenges for the application of copyright law is raised by the phenomenon of peer-to-peer file sharing. Users consider P2P file sharing to be a new utility of enjoyment, while copyright holders insist that the new application be placed under marketable exploitation.

Indeed, "[r]ights holders perceive these and other offerings as fairly within the exclusive rights of copyright and directed toward markets that are reasonable and likely to be developed."[465] As Michael Meurer points out:

> In economic terms, a pirate is a competing supplier of the copyright holder, while parties engaged in sharing (whether authorized or not) should be thought of as a coalition of buyers. As the boundary between public and private distribution becomes blurred unauthorized sharing starts to look like piracy, and the organizer of a coalition of buyers begins to look like a competing supplier.[466]

New digital technology has posted new challenges for the reformulation of the new interest pattern in copyright law. As innumerable unlicensed transmissions and P2P file sharing of copyrighted works by end users increasingly constitute a formidable threat to the commercial exploitation of works by copyright holders,[467] copyright industries have lobbied legislators to expand their exclusive rights into digital networks as a response. Moreover, copyright holders have deployed TPMs to control access and block unauthorized copying of their works while forbiding users to circumvent them.[468] The perfect control offered by TPMs has raised fierce debates on the status of private use in the digital environment.[469]

3.2 Efforts to Harmonize TPMs and Private Use

In this section, I intend to introduce some positive efforts to harmonize TPMs and private use in the digital environment, either from the legislative perspective or from judicial practice. The selected measures or experiences can arguably serve as examples for other countries to deal with the conflict between TPMs protection and private use.

---

465 Pamela Samuelson et al., *supra* note 171, at 54.

466 Meurer, *supra* note 454, at 852.

467 See RIAA, *supra* note 39.

468 See e.g., Article 11 of the WCT, and Article 18 of the WPPT.

469 See *infra* sections 3.3 & 3.4.

3.2.1 Prerequisite of TPMs Protection in the 1996 WIPO "Internet Treaties"

In order to regulate the new type of utilizing copyrighted works under the digital environment, the two WIPO "Internet Treaties" of 1996 – the WCT and the WPPT – contain "the right of making available" to cover the offering and diffusion of works, performances, and phonograms on the internet and similar networks.[470]

Although on-demand[471] and other internet uses are covered by the making available right under the WCT and the WPPT, member states could implement this right in various ways at the domestic level.[472] For instance, China has chosen the way of granting a new exclusive right – the information network transmission right – to copyright holders to implement the WIPO Treaties' obligation.[473]

Besides confirming the new right of making available, the WCT and the WPPT also contain obligations concerning technological measures and rights management information.[474] "Unlike minimum rights, they do not grant any direct protection against unauthorized uses, but aim at indirectly protecting the minimum rights by flanking legal measures and thereby indirectly serve the enforcement of the minimum rights."[475]

---

470 Under the WCT, the right of making available has been made a part of the right of communication to the public. See Article 8 of the WCT (right of communication to the public). Under the WPPT, however, it has been drafted as a right separate from the right of communication to the public, see Article 10 (right of making available of fixed performances), Article 14 (right of making available of phonograms), and Article 15 (right to remuneration for Broadcasting and Communication to the Public).

471 On-demand use means making the works available to the public in such a way that members of the public may access these works from a place and at a time individually chosen by them. See Article 8 of the WCT.

472 The WIPO Treaties allow implementation of the making available right by any suitable right, which is so-called "umbrella-solution". For example, the member states could implement the making available right by a right of broadcasting, distribution, communication to the public or others. See MIHÁLY FICSOR, THE LAW OF COPYRIGHT AND THE INTERNET: THE 1996 WIPO TREATIES, THEIR INTERPRETATION AND IMPLEMENTATION C.8.06-C.8.23 (Oxford: Oxford University Press 2002) (referring to different ways of implementation at national level).

473 See Article 10 (12) of the 2010 Chinese Copyright Law.

474 See Articles 11 and 12 of the WCT, and Articles 18 and 19 of the WPPT.

475 SILKE LEWINSKI, INTERNATIONAL COPYRIGHT LAW AND POLICY 462 (NY: Oxford University Press 2008).

Notwithstanding the fact that it is not compulsory for the member states to impose the use of TPMs and rather the decision is left at the discretion of rights holders to decide, the Contracting Parties are required to provide adequate legal protection and effective legal remedies against the circumvention of "effective" technological measures that are used by copyright holders.[476] However, the meaning of "circumvention" is ambiguous. It is unclear whether it indicates behavior which directly circumvents the technological measures, or behavior which indirectly provides the services that are used to circumvent, or both.

Notably, the technical measures protected by the WIPO Internet Treaties must be employed to protect uses by right holders in connection with the exercise of their rights in respect of their works, which are not authorized by the copyright holders or permitted by law.[477] This prerequisite of TPMs protection emphasizes two considerations.

First, the purpose of employing TPMs is to protect the uses in connection with the exercise of copyright under the Treaties or the Berne Convention. It is very interesting that the Treaties use the phrase of "in connection with" the rights under the Treaties rather than directly indicating the rights under the Treaties. This artful expression leaves some leeway for Contracting Parties to provide greater protection for TPMs. Take the access control measures as an example. Principally there is no "right of access" under the Berne Convention or the Treaties, but the access control measures could be protected because it is in connection with a protected copyright, such as the reproduction right, distribution right, and communication right.

Second, technological measures protected by the Treaties must be employed to protect uses which are not authorized by the copyright holders or permitted by law. The uses permitted by the law particularly indicate the copyright limitation or exemption. For example, if works in the public domain are protected by technical measures, the circumvention of TPMs does not violate the Treaties. Similarly, regarding private use and other permitted uses by law, the circumvention of TPMs also does not violate the Treaties.

In fact, this prerequisite of TPMs protection is intended to keep a balance between copyright protection against circumvention and the limitation system for the sake of the public interest. However, it is highly difficult, as required by the Treaties, to find the right balance between effective protection and respect for limitations and exceptions, because devices cannot distinguish between

476 See Article 11 of the WCT, and Article 18 of the WPPT.

477 See *id.*

permitted or non-permitted uses.[478] Consequently, the Treaties do not provide concrete prescriptions regarding the harmonization between TPMs protection and copyright limitation. Member states could implement in various manners as long as this basic precondition is followed.

3.2.2 Saving Clauses on TPMs Protection in the Digital Millennium Copyright Act of 1998

The DMCA divides technological measures into two categories: measures that prevent unauthorized access to a copyrighted work, and measures that prevent the unauthorized exercise of the exclusive rights of a copyright holder. As to the measures that prevent unauthorized access to a copyrighted work, it not only prohibits the making or selling of devices or services that are used to circumvent, but also the act of circumvention itself. While as to the measures that prevent the unauthorized use of a copyrighted work, it only prohibits the making or selling of devices or services that are used to circumvent.[479]

This distinction is said to assure that the public can make fair use of copyrighted works, as illustrated in the explanation of the DMCA by the US Copyright Office:

> This distinction was employed to assure that the public will have the continued ability to make fair use of copyrighted works. Since copying of a work may be a fair use under appropriate circumstances, section 1201 does not prohibit the act of circumventing a technological measure that prevents copying. By contrast, since the fair use doctrine is not a defense to the act of gaining unauthorized access to a work, the act of circumventing a technological measure in order to gain access is prohibited.[480]

At the same time, the DMCA contains a general saving clause which explicitly indicates that nothing in section 1201 affects rights, remedies, limitations or defenses to copyright infringement, including fair use.[481] It seems that the protection of TPMs is undoubtedly subject to the fair use exemption. However, in the light of the above interpretation by the US Copyright Office, the copyright

478 See SAM RICKETSON & JANE GINSBURG, INTERNATIONAL COPYRIGHT AND NEIGHBOURING RIGHTS: THE BERNE CONVENTION AND BEYOND 15.19 (NY: Oxford University Press 2006).

479 See 17 U.S.C. § 1201.

480 The Digital Millennium Copyright Act of 1998, U.S. Copyright Office Summary, Dec. 1998, p4.

481 See 17 U.S.C. §1201(c)(1).

limitations only adapt to the copyright protection measures, but not the access control measures.

The discriminating protection pursuant to the different types of TPMs generates a complicated theoretical problem. In respect of the logic of dividing TPMs into access control measures and copyright protection measures, the US Congress does not recognize access control measures protecting copyright. The "copyright protection measure" means a measure that in the ordinary course of its operation, prevents, restricts, or otherwise limits the exercise of a right of a copyright owner.[482] Obviously, this kind of TPMs protects the exclusive rights of the copyright holder. In theory, since these exclusive rights are subject to the fair use exemption, the technological measures which protect copyright are also restricted by the same limitations. Therefore, given the fact that the circumvention of copyright protection measures might be the fair use of the works, it should not be generally forbidden.

In contrast, the definition of "access control measure" is that which, in the ordinary course of its operation, requires the application of information, or a process or a treatment, with the authority of the copyright owner, to gain access to the work.[483] Because this restriction of the access to works is not the traditional exclusive right of copyright holders, it says nothing of the copyright limitations. Since the access control measures are not subject to the fair use exemption, and the fair use doctrine is not a defense to the act of gaining unauthorized access to works, the circumvention of a technological measure in order to gain access is prohibited by the DMCA.

In other words, the US copyright law does not admit access control measures which protect copyright, but the fact is that protecting the access control measure itself through copyright law is obviously intended to protect copyright. This leads to a surprisingly strange logic of legislation – "to protect the TPMs which does not protect copyright" equals to "protect copyright".[484] Moreover, if the copyright law does not authorize the right holder the exclusive right to prohibit others' unauthorized access to a work, then what is the legal justification of a right holder to employ the access control measures?

In order to solve this flaw in the logic, some scholars advocate authorizing an "access right" to the author since the exploitation of the works shifts from having copies to directly experiencing the content of the work.[485] "As we move to an access-based world of distribution of copyrighted works, a copyright

482 See 17 U.S.C. §1201(b)(2)(B).

483 See 17 U.S.C. §1201(a)(3)(B).

484 See Wang Qian, *supra* note 195, at 5.

485 See e.g., Ginsburg, *supra* note 43.

system that neglected access controls would make copyright illusory, and in the long run it would disserve consumers."[486] In this way, the purpose of the access control measures is to protect the exclusive right of copyright holders – the access right – and then it would be reasonable to protect the access control measures by the copyright law.

However, a number of commentators have provided a fierce critique of TPMs.[487] The regulation of TPMs in the DMCA has been criticized as being deployed to undermine the fair use defense, and to disrupt the traditional balance between copyright owner rights and user privileges in copyright law.[488] Armed with technological measures and the right to prevent circumvention, copyright holders can control access to a work, and require payment each time per use, even with respect to a work that is no longer protected by copyright law. As the "fair use" is becoming "fared use",[489] the DMCA lays the legal foundation for a regime of "universal pay-per-use and *de facto* perpetual protection."[490]

3.2.3 Judicial Restraint on the Application of Anticircumvention Rules: MGE UPS Systems Inc. v. GE Consumer and Industrial Inc.

The WCT and the WPPT indicate that the protected TPMs shall be "effective" and "in connection with the exercise of" copyrights.[491] However, the Treaties offer no definitions whatsoever. In this section, I intend to explore a conservative application of anticircumvention rules according to the

486 *Id.* at 124.

487 See e.g., LAWRENCE LESSIG, FREE CULTURE: HOW BIG MEDIA USES TECHNOLOGY AND LAW TO LOCKDOWN CULTURE AND CONTROL CREATIVITY (NY: Penguin 2004); Fred Lohmann, *Measuring the Digital Millennium Copyright Act Against the Darknet: Implication for the Regulation of Technological Protection Measures*, 24 LOYOLA OF LOS ANGELES ENTERTAINMENT LAW REVIEW 635 (2004); Litman, *supra* note 4; Koren, *supra* note 48.

488 See Lawrence Lessig, *id.* at 160.

489 See Tom Bell, *Fair Use vs. Fared Use: The Impact of Automated Rights Management on Copyright's Fair Use Doctrine*, 76 N. C. L. REV. 557 (1998).

490 David Nimmer, *How Much Solicitude for Fair Use Is There in the Anti-Circumvention Provision of the Digital Millennium Copyright Act?* in THE COMMODIFICATION OF INFORMATION 193, 220 (Niva Elkin-Koren & Neil Weinstock Netanel eds., The Hague: Kluwer Law International 2002).

491 See Article 11 of the WCT, and Article 18 of the WPPT.

"effectiveness" and "connection with" copyright requirements in a recent US case – *MGE UPS Systems Inc. v. GE Consumer and Industrial Inc.*[492]

MGE manufactures uninterruptible power supply (UPS) machines which are used during periods of power outages to provide power to critical operation systems. UPS machines require the use of MGE's copyrighted software during servicing. The software demands connection of an external hardware security key – a so called "dongle". Each dongle has an expiration date, a maximum number of uses, and a unique password. A number of software hackers published information on the internet disclosing how to defeat the "dongle".

PMI is a critical power service company servicing customers including MGE's UPS machines. PMI initially subcontracted MGE to perform the dongle on MGE UPS machines. Afterwards, a group of PMI employees obtained hacked MGE software. In 2001 GE acquired PMI.

In December 2004, MGE sued GE/PMI for copyright infringement, misappropriation of trade secrets, unfair competition, and violation of the DMCA. The US District Court for the Northern District of Texas awarded MGE damages on a jury verdict, post-judgment interest, a permanent injunction against GE/PMI, impoundment of infringing materials, taxable costs, and attorney's fee. However, the District Court dismissed the DMCA claim. The case was appealed to the US Court of Appeals for the 5th Circuit.

On July 20, 2010, US Court of Appeals for the 5th Circuit released a decision interpreting Section 1201(a) of the DMCA in *MGE UPS Systems Inc. v. GE Consumer and Industrial Inc*. In its decision, the US Court of Appeals for the 5th Circuit adopted a restrictive judicial interpretation on the DMCA's anticircumvention rules.

First, the court opposed an excessively broad definition of "access". Otherwise, the broad interpretation would permit liability under § 1201(a) for accessing a work simply to view it or to use it within the purview of "fair use" permitted under the copyright law. The court indicated that "[w]ithout showing a link between 'access' and 'protection' of copyrighted work, Digital Millennium Copyright Act's (DMCA) anti-circumvention provision does not apply."[493] That is to say, the DMCA only prohibits the access that would violate or impinge on a right that the copyright law protects. Merely bypassing a technological protection that restricts a user from viewing or using a work is insufficient to trigger the DMCA's anticircumvention rules.

---

492 See *MGE UPS Systems Inc. v. GE Consumer and Industrial Inc.*, 2010 WL 2820006 (C.A.5 (Tex.)).

493 See *Chamberlain Group, Inc. v. Skylink Techs., Inc.*, 381 F.3d 1178, 1204 (Fed.Cir.2004).

Second, the DMCA's anticircumvention provision only targets the circumvention itself, but does not apply to the use of copyrighted works after the technological protection measure has been circumvented.[494]

3.2.4 Copyright Limitations as Users' Right: CCH Canadian Limited v. Law Society of Upper Canada

In *CCH Canadian Limited v. Law Society of Upper Canada*,[495] the Supreme Court of Canada established the boundary of fair dealing in Canadian copyright law. The decision has been applauded as bringing forth a truly fundamental shift in the way Canadian copyright law is to be understood and applied.[496] The decision is especially significant, because of "its affirmation of 'user rights' as a concept integral to copyright law. In the Court's eyes, user rights are as central to copyright law as author rights. CCH thus affirms the irreducible centrality of the public domain in Canadian copyright law jurisprudence".[497]

In the decision, the Supreme Court provides an understanding of fair dealing, and recognizes it as an integral part of the copyright law rather than merely a negative defense. Moreover, the Supreme Court explicitly indicates that the fair dealing and other exceptions in the copyright law are users' rights. As the following quote illustrates:

> Before reviewing the scope of the fair dealing exception under the Copyright Act, it is important to clarify some general considerations about exceptions to copyright infringement. Procedurally, a defendant is required to prove that his or her dealing with a work has been fair; however, the fair dealing exception is perhaps more properly understood as an integral part of the Copyright Act than simply a defence. Any act falling within the fair dealing exception will not be an infringement of copyright. The fair dealing exception, like other exceptions in the Copyright Act, is

494 See *Universal City Studios, Inc. v. Corley*, 273 F.3d 429, 443 (2d Cir.2001) (pointing out that "the DMCA targets the circumvention of digital walls guarding copyrighted material (and trafficking in circumvention tools), but does not concern itself with the use of those materials after circumvention has occurred.").

495 See *CCH Canadian Limited v. Law Society of Upper Canada*, [2004] 1 S.C.R. 339, 2004 SCC 13.

496 See e.g., Teresa Scassa, *Recalibrating Copyright Law? A Comment on the Supreme Court of Canada's Decision in CCH Canadian Ltd. v. Law Society of Upper Canada*, 3 CJLT 89 (2004).

497 Abraham Drassinower, *Taking User Rights Seriously*, University of Toronto, Faculty of Law, Legal Studies Series, Research Paper No. 09-05, Nov. 2005.

> a user's right. In order to maintain the proper balance between the rights of a copyright owner and users' interests, it must not be interpreted restrictively.[498]

The statement of "a user's right" by the Canadian Supreme Court clearly shows its position regarding copyright limitations. The Court cites David Vaver's argument indicating that "User rights are not just loopholes. Both owner rights and user rights should therefore be given the fair and balanced reading that befits remedial legislation."[499] This positive right could empower private users to protect their interests in cases of copyright limitations against undue encroachment via TPMs or any other means employed by copyright holders.

3.3 Assumption I: "Unauthorized Copying is Piracy"

Given the fact that the conventional distinction between enjoyment and exploitation has become blurred in the digital environment, some commentators have suggested that private copying should no longer be tolerated and any unauthorized copying is piracy. In this section, I intend to explore their main arguments.

3.3.1 The Market Failure Justification of Private Copying Loses Its Importance in the Digital Environment

Traditionally, the market failure consideration has been an important justification for private use. When the public goods characteristics of intellectual creations make the bargaining between copyright holders and potential users very costly, or make them unable to enforce their rights effectively against unauthorized uses, market failure is said to occur. As an alternative, private use is permitted. However, in the digital environment, the market failure justification for private copying has gradually lost its importance and this has created a challenge for the maintenance of the private copying exemption.

This section begins with an analysis of the two public goods characteristics of intellectual creations. Then, I will explore the market failure justification for private copying as well as its challenges in the digital environment.

---

498 *CCH Canadian Limited v. Law Society of Upper Canada*, [2004] 1 S.C.R. 339, 2004 SCC 13, Para. 48.

499 *Id.* citing DAVID VAVER, COPYRIGHT LAW 171 (Toronto: Irwin Law 2000).

3.3.1.1 The Public Goods Characteristics of Intellectual Creations

Public goods have two inherent qualities – non-rivalry and non-excludability[500] – which separate them from private goods. Non-rival means that a good's position or consumption by one person does not reduce the quantity of the same good available to other people.[501] In other words, "the possession and consumption of 'public goods' is deemed 'non-rival' because one person's possession or consumption of the property does not typically affect the ability of others to simultaneously possess or derive utility from the property."[502] For example, watching a movie on TV, or listening to a song on radio does not prevent others from consuming the same movie or song at the same time.

A good is non-excludable when, once it has been produced and released to the public, it becomes impossible to exclude an individual from consuming that good even if the individual did not contribute to the cost of producing it.[503] For example, once a book is published and sold, its publisher can not stop the same book from being borrowed or read by several people who have not paid.

In comparison, a private good is characterized as "rival" and "excludable". The possession or consumption of the good by one person does deprive other people of the ability to simultaneously possess or consume the good.[504] For example, eating food, driving a car, or sleeping on a bed does exclude others from consuming or using the same good.

Accordingly, public goods and private goods are different in regards to control of access. "In general, it is typically relatively easy to restrict access to private goods.... such as attaching a lock, or merely retaining physical possession of the property. But it is much harder to control and regulate access to public goods."[505]

Intellectual creations are often regarded as public goods because they are non-rival and non-excludable.[506] The "distinctive characteristics of most

---

500 See Shubha Ghosh, *Pills, Patents, and Power: State Creation of Gray Markets as a Limit on Patent Rights*, 53 FLA. L. REV. 789, 797 (2001).

501 Peter Eckersley, *Virtual Market for Virtual Goods: The Mirror Image of Digital Copyright?* 18 HARV. J. L. & TECH. 85, 117 (2004).

502 Lee Bollinger, *Book Review: Protect This Work of Expression: Clarifying the Unique Economics of Intellectual Property Rights, Steal This Idea: Intellectual Property Rights and the Corporate Confiscation of Creativity by Michael Perelman*, 44 SANTA CLARA L. REVIEW 1287, 1290-91(2004).

503 See Eckersley, *supra* note 501, at 117-18.

504 See Bollinger, *supra* note 502, at 1290.

505 YIJUN TIAN, RE-THINKING INTELLECTUAL PROPERTY: THE POLITICAL ECONOMY OF COPYRIGHT IN THE DIGITAL ERA 71 (NY: Routledge-Cavendish 2009).

506 See Peter Drahos, *The Regulation of Public Goods*, 7 J. INT. ECON. L. 321, 321 (2004).

intellectual products are that they are easily replicated and that enjoyment of them by one person does not prevent enjoyment of them by other persons".[507] Thus, "[u]nlike private property, it is difficult to control the use and dissemination of intellectual property once it is released".[508]

3.3.1.2 Transaction Costs of Forbidding Private Copying: Market Failure Consideration

"When approaching IPR issues, Utilitarianism theorists often begin their discussion with an analysis of the nature of IP products as 'public goods', and then focus on cost benefit analysis when discussing the social welfare effects of IP law implementation."[509] From an economics point of view, the public goods characteristics of intellectual creation induce concerns of underutilization and underproduction.[510]

The problem of underutilization is generated by the non-rivalry characteristic of intellectual creation. Since many people can consume public goods without depleting them, it is very difficult or costly to identify who has not paid for access and to prevent potential free riders from using those goods.[511] As Giuseppe Mazziotti indicates:

> In economic terms, non-rivalry implies that the marginal cost of serving an additional consumer is zero. If marginal cost is assumed to be zero for non-rival goods, then property rights permitting royalties to be charged to additional consumers lead to an inevitable "deadweight loss" to society. Indeed, if the good is made available at a price that is higher than the marginal cost (where the marginal cost is zero), only consumers who are willing to pay the price set by the right holder (that is to say, a price higher than zero) are allowed to benefit from the good. Consumers whose willingness to pay is higher than zero but lower than the price set

---

507 William Fisher, *Theories of Intellectual Property*, in NEW ESSAYS IN THE LEGAL AND POLITICAL THEORY OF PROPERTY 168 (Stephen Munzer ed., 2001), available at http://cyber.law.harvard.edu/people/tfisher/iptheory.pdf (last visited Oct. 6, 2010).

508 Tracy R. Lewis & Dennis A. Yao, *Some Reflections on the Antitrust Treatment of Intellectual Property*, 63 ANTITRUST L. J. 603, 606 (1995).

509 TIAN, *supra* note 505, at 70.

510 See GIUSEPPE MAZZIOTTI, EU DIGITAL COPYRIGHT LAW AND THE END-USER 15 (Heidelberg: Springer 2008).

511 See Kenneth J. Arrow, *Economic Welfare and the Allocation of Resources for Inventions*, in THE RATE AND DIRECTION OF INVENTIVE ACTIVITY: ECONOMIC AND SOCIAL FACTORS 614-16 (Richard R. Nelson ed, Princeton: Princeton University Press 1962).

> by the right holder are thereby excluded from consumption, even if they would have consumed the good at no cost to anyone.[512]

Another problem of underproduction stems from the characteristic of non-excludability. Given the fact that it is impossible to exclude others from using intellectual products, free riders would reduce the rewards of creativity and make creators unable to recoup their costs. When the investment in intellectual products becomes "uniquely risky and unappealing", there will be insufficient economic incentive for creators to engage in creativity.[513] As a result, underproduction will emerge as a result of free riders.[514]

Consequently, IP laws are recognized as an effort to cure a form of market failure stemming from the public goods characteristics of intellectual creations.[515] In order to overcome the problem of underutilization, IP laws grant right holders exclusive rights only for a limited time. After the legislative duration expires, intellectual creations enter into the public domain, and are available to everyone free of charge.

Meanwhile, by making the intellectual products excludable for a limited time and scope, IP laws allow right holders to charge royalties in order to solve the problem of underproduction.[516] As Peter Drahos observes, "IP rights are essentially means of permitting exclusive use to knowledge in order to encourage its further development".[517] By doing so, IP laws attempt to "strike a balance between the incentive to create and innovate, and the diffusion of the results obtained".[518]

However, the copyright market still might not always function adequately. Although copyright law has provided a means for excluding free riders so as to

---

512 MAZZIOTTI, *supra* note 510, at 16.

513 See Bollinger, *supra* note 502, at 1291 (indicating that "in the face of a product whose accessibility is not easily regulated and whose supply is largely unaffected by consumption, producers have few incentives to invest in products with 'public goods' characteristics".).

514 See Michael Carrier, *Unraveling the Patent-Antitrust Paradox*, 150 U. PA. L. REV. 761, 767 (2002).

515 See Gordon, *supra* note 33, at 1610.

516 See Bollinger, *supra* note 502, at 1291 (stating that "one of the predominant justification of the IP laws is to avoid the 'market failure' (or the unacceptably low level of production as a matter of public policy) that would otherwise occur due to the insufficient incentives to create goods that have little profit-making potential to their creators but which have potentially high social and economic value to society".).

517 Drahos, *supra* note 506, at 322.

518 François Lévêque & Yann Ménière, *The Economics of Patens and Copyright*, 5 (The Berkeley Electronic Press July 2004), available at http://www.bepress.com/leveque/ (last visited Oct. 11, 2010).

cure the public goods problem, at times high transaction costs may make bargaining between copyright owners and potential users of copyrighted material impossible or prohibitively costly and copyright owners may be unable in practice to enforce their rights effectively against unauthorized users. Market failures persist as a result.[519] In response, the doctrine of fair use has been introduced as a cure "when flaws in the market might make reliance on the judiciary's own analysis of social benefit appropriate".[520] Accordingly, private copying is permitted for practical reasons, namely it being too expensive to prohibit due to market failure.

3.3.1.3 No Market Failure, No Private Copying Exemption

In the analogue world, IP laws create artificial excludability in order to solve public goods problem.[521] This artificial excludability never amounts to perfect control over copyrighted works but is limited in time and also in scope. "The non-rival nature of information ensures that the excludability provided by copyright, even in the most individualistic systems, does not amount to perfect control over protected work."[522]

By contrast, in the digital environment, right holders can implement perfect control over their copyrighted works by employing TPMs and the right to prevent circumvention.[523] By making the digital content excludable and forcing people to pay for access, the TPMs framework functions in the same way as a ticket system, which serves as an exclusion tool that has "no other utility or additional value to the consumer than making him pay".[524]

---

519 See Gordon, *supra* note 33, at 1613.

520 See *id.* at 1614.

521 See Mark A. Lemley, *Property, Intellectual Property, and Free Riding*, 83 TEX. L. REV. 1031, 1055 (2005) (stating that IP is "a conscious decision to create scarcity in a type of good in which it is ordinarily absent in order to artificially boost the economic returns to innovation".).

522 MAZZIOTTI, *supra* note 510, at 17.

523 *Cf.* Merges expresses skepticism about perfect excludability: "even the strongest digital rights management system one might contemplate falls far short of total control over the information it protects from unwanted copying [...] there are many manifestations of digital content that cannot be effectively controlled by a right holder." See Robert P. Merges, *Compulsory Licensing vs. the Three "Golden Oldies"*, POLICY ANALYSIS, No. 508, Jan. 15, 2004, p.5, available at http://www.cato.org/pubs/pas/pa508.pdf (last visited Oct. 11, 2010).

524 See Olivier Bomsel & Anne-Gaëlle Geffroy, *Economic Analysis of Digital Rights Management Systems (DRMs)*, MediaNet Project Paper, reviewed version of Feb.10,

If private copying was once lawful solely because of enforcement difficulties due to market failure, the easy and low cost of enforcement by TPMs against unauthorized use would remove the conventional market failure justification for private copying.[525] As Jane Ginsburg suggests, "if the justification for private copying is 'everyone does it,' then another solution is to make it harder for everyone to do it."[526]

3.3.2 From de minimas to Multiple: Private Copying as Competitive Exploitation in the Digital Environment

In the analog world, private copying through handwriting or typewriting is generally perceived as non-infringing because it is *de minimas*.[527] However, with the development of digital technology, private copying by individuals has become easy, costless, and large-scale. In addition, there is almost no quality derogation during the reproduction process. Some commentators argue that since private copying is no longer *de minimas* but large-scale, and increasingly constitutes a formidable threat to right holders' exploitation, unauthorized private copying should no longer be permitted in the digital environment.

---

p.14, 2005, available at http://www.cerna.ensmp.fr/Documents/OB-AGG-EtudeDRM.pdf (last visited Oct. 12, 2010).

525 A counter argument would be that there is still market failure in cyberspace. For example, bargaining breakdown potentially happens in parodies and criticisms, where right holders are likely to refuse permission to users. In the famous "Wind Done Gone" case (*Suntrust Bank v. Houghton Mifflin Co.*, 252 F.3d 1165 (11th Cir.2001)), the author Alice Randall retold the story of Margaret Mitchell's *Gone with the Wind* with new characters and new interpretations of events. Because of the plot of homosexuality and miscegenation, the original author sued the derivative work for copyright infringement. In the end, the work was regarded as parody and protected by fair use. However, the case shows that this kind of secondary work could hardly, if not never, get the authorization from the original author.

526 Ginsburg, *supra* note 8, at 19 (the author further suggests that "first, that works for which the author seeks payment, and whose subsequent copying the author wishes to limit, will not be made available over open networks. Second, these works will not be disseminated without technological controls against unauthorized access or further copying.").

527 See Leval, *supra* note 32, at 1457-58.

3.3.2.1Private Copying as Displaced Sales

Although the exact legal status of hand-copying for private use is more vague than clear, early laws on copyright were generally not concerned with the small-scale and hand-made reproduction that occurred in the private sphere. As Melville Nimmer and David Nimmer point out "[t]here is no reported case on the question of whether a single handwritten copy of all or substantially all of a book or other protected work made for the copier's own private use is an infringement or fair use."[528] However, in the *Williams & Wilkins Co. v. United States*, it is stated that "it is almost unanimously accepted that a scholar can make a handwritten copy of an entire copyrighted article for his own use".[529] In any case, "it was a question no one would be interested to litigate".[530]

With the rise of photocopying and analogue recording equipment in the 1950s,[531] individuals were able to make relatively easy and cheap copies of works. In order to compensate the copyright holder's economic remuneration affected by private copying, the levy system was introduced to harmonize the competing interests between the right holders' exploitation and the users' enjoyment. Consequently, private copying is legally permitted in exchange for a levy paid on copying media.

In the newly digital environment, the previously inconspicuous private copying has become highly suspect. At present, "[t]he growing economic importance of private copying is no mirage."[532] "Private copying can also have commercial consequence. As ...technologies ...has made copying possible where it was impossible before, the risk has grown that 'private' copies will displace the retail sales and rentals of the authorized copies from which ... [copyright holders] earn their revenues."[533]

There has been no consensus on this observation. Some other commentators express their disagreement. The argument that private copying constitutes displaced sales and thus damages right holders' economic profits, "presupposes the existence of a right to profit on private use. Such a right, however, does not clearly exist...the assertions that private use causes economic damages

528 NIMMER ON COPYRIGHT, *supra* note 155, §13.05[E].

529 *Williams & Wilkins Co. v. United States*, 487 F. 2d 1345, 1350 (Ct. Cl. 1973).

530 KAPLAN, *supra* note 22, at 68.

531 See GOLDSTEIN, *supra* note 233, at 106 ("In 1983, less than ...9 percent of the American homes with television also had videocassette recorders; by 1991, the proportion was 71 percent. In 1990, the average American household had more than two audiotape recorders capable of making copies.").

532 *Id.*

533 *Id.*

presupposes the transgression of a legal right, and cannot, therefore, be the foundation of that right."[534]

3.3.2.2 Accumulative Noncommercial Use Equals Commercial Use

Many a little makes a mickle. Multiple noncommercial copying by individuals "can combine to make something that seems commercial in scale and threatens to undermine copyright owners' opportunities to explore their works commercially".[535] Even if a single private copying does not constitute a replaced sale, the accumulative effect of a large amount of noncommercial private copying might equal commercial use, or at least have an effect on the potential market.

This observation is not new. When photocopying arose, the validity of private copying came under suspicion. As the Office of Technology Assessment (OTA) of the US Congress indicated:

> To decide the issue of reprography, the courts drew an analogy between photocopying and making a handwritten copy of the work for private consumption. Photocopying was regarded as merely a more efficient way of doing the same thing as hand copying, and as such, was not considered to be an infringement. Although similar in principle, the analogy between hand and photocopying soon breaks down, however. For, precisely because photocopying is more efficient than hand copying, its potential to cause economic harm is much greater. This is the problem of harm in the aggregate, a problem that becomes more troublesome as technology continues to improve the efficiency of private copying.[536]

The "problem of harm in the aggregate" observation was shared by Melville Nimmer. As he points out, "[f]ew people hand copy, but millions find machine copying economical and convenient.... [The argument] fails to take into account the true economic effect when thousands of individual decisions are aggregated."[537]

Similar consideration could also be found in *Sony Corp. v. Universal City Studios*. The Supreme Court stated that, "[a] challenge to a non-commercial use of a copyrighted work requires proof either that the particular use is harmful, or that if it should become *widespread*, it would adversely affect the potential

---

534 US CONGRESS, OFFICE OF TECHNOLOGY ASSESSMENT, INTELLECTUAL PROPERTY RIGHTS IN AN AGE OF ELECTRONICS AND INFORMATION, OTA-CIT-302, 198 (Washington, DC: US Government Printing Office, April 1986).

535 Litman, *supra* note 4, at 1913.

536 US CONGRESS, *supra* note 534, at 197.

537 Melville B. Nimmer, *New Technology and the Law of Copyright: Reprography and Computers*, 15 UCLA L. REV. 931, 951 (1968).

market for the copyrighted work."[538] In other words, if the accumulative effect impacts the potential market of exploitation, the non-commercial characteristic of private copying would be denied.

Moreover, from a "diachronic" perspective, digital private copies compete with the market sales. In the pre-digital era, private copies cannot fully substitute for the original because of the quality derogation during the reproduction process. In the digital era, however, the copies remain almost the same with the original. Therefore, if the private user makes a digital copy, it can be used repeatedly and theoretically for ever.

In this sense, private users become media of themselves for their future use. This would greatly harm copyright holders' benefit through authorizing commercial use. Thus, in order to keep a balance in the competition between authorized distribution of contents by commercial distributor and personal copy and enjoyment of contents by individuals, it is necessary for copyright law to limit not only the numbers of private copies, but also the time scale of the use of copies.

In a word, private copying is no longer inoffensive *de minimas*, but is carried out on a large scale and constitutes displaced scales and competitive exploitation. Even if a single private copying does not constitute a replaced sale, the accumulative effect of a large amount of noncommercial private copying might equal to commercial use, or at least have an effect on the potential market. Unauthorized copying should therefore no longer be tolerated in the digital environment.

3.3.3 The Necessity of Access Control by TPMs

In respect of the innumerable instances of the unauthorized copying, transmission and file sharing of copyrighted works, some commentators argue that access control by TPMs is important and necessary as the exploitation of works has shifted from having copies to experiencing works.[539] Moreover, by virtue of TPMs access control and price discrimination, users only need to pay certain fees depending on the various uses of works. This new business model, some argue, would be more reasonable and affordable for both right holders and users.

---

538 *Sony Corp. v. Universal City Studios*, 464 U.S. 417 (1984) [emphasis added].

539 See Ginsburg, *supra* note 43.

3.3.3.1 From Having Copies to Experiencing Works: The Shifting Exploitation of Works

In the pre-digital era, private copying served an important role for users' enjoyment. Users needed physical copies in order to enjoy copyrighted works. For example, in *Sony v. Universal Studios*, users used VCRs to record broadcast programs off the air. Private copying was necessary for later viewing at home.[540]

In comparison, in the digital environment, physical copies might no longer be needed for users to enjoy copyrighted works. As Jane Ginsburg suggests:

> In the impending era of digital access, we will be able to download anything (at least temporarily), whenever and wherever we want. As a result, we will no longer need hard copies to enjoy the work; indeed, in a world of access conditioned on nonretention of digital copies, we will be able to summon up the work at any time, but we may not be able to have our own copy.[541]

With instant access through digital networks to various works of literature, music, movies, and other digital version of copyrighted works, users can enjoy the works at any time and at any place as long as there is an internet connection. Accordingly, hard copies need no longer be retained in hard disk or other mediums.

However, without sufficient assurance, copyright holders will hesitate to make their works available for instantaneous online distribution. They need to be confident that the "experiencing" copy will not change into the unauthorized "having" copy or "sharing" copy.[542] "[W]hen the exploitation of works shifts from having copies to directly experiencing the content of the work, the author's ability to control access becomes crucial."[543]

In response, "in the digital environment, the 'exclusive Right' that the U.S. Constitution authorizes Congress to secure to authors is not only a 'copy'-right, but an access right".[544] The access right is part of the copyright regime in the 21st century. It does not and should not supplant copyright but should be viewed as being an integral part of copyright.[545]

---

540 See *Sony Corp. of America v. Universal City Studio Inc.* 464 U.S. 417 (1984).

541 Jane C. Ginsburg, *From Having Copies to Experiencing Works: The Development of an Access Right in U.S. Copyright Law*, in US INTELLECTUAL PROPERTY LAW AND POLICY 41 (Hugh Hansen ed., UK: Edward Elgar Publishing Limited 2006).

542 See Senate Comm. On the Judiciary, The Digital Millennium Copyright Act of 1998, S. Rep. No. 105-90, at 8 (1998).

543 Ginsburg, *supra* note 43, at 115-16.

544 See Ginsburg, *supra* note 541, at 42.

545 See *id.*

More specifically, in order to provide economic incentives to creativity and protect the author's intellectual fruits, it is necessary for the access right to prohibit unauthorized acquisition of copies of copyrighted works. Meanwhile, the access right should also be subject to copyright limitations, including fair use. However, the fair use defenses to an access right should be carefully considered and adapted to digital online distribution and will not necessarily be fully consistent with fair use defenses to traditional copyright.[546]

3.3.3.2 Price Discrimination Depending on Various Uses of Works

Price discrimination,[547] or value-based pricing, is an increasingly important business model that seeks to harmonize the competing interests between digital content providers and users. "Roughly speaking, price discrimination is a tactic by which a seller charges a high price to high valuation users and a low price to low valuation users."[548] In order to satisfy multiple consumption demands and maximize profit, price discrimination is widely adopted by various businesses. For example, a few years ago, the computer manufacturer Dell sold its 512 MB memory module to diverse clients at different prices: $289.99 for a large business, $266.21 for a government agency, $275.49 for a home, and $ 246.49 for a small business.[549]

546 See *id.* at 42-58.

547 Historically, Arthur Cecil Pigou divided price discrimination into three types. The first type is perfect price discrimination, meaning that each buyer paid 100% of the subjective value of the goods purchased, and prices were based on the buyers' identities. The second category indicates a menu of options, such as a quantity discount, and letting buyers choose what to buy. The third type refers to an imperfect form of the first type. Quoting JEAN TIROLE, THE THEORY OF INDUSTRIAL ORGANIZATION 135 (Massachusetts Institute of Technology 1988); *cf.* R. Preston McAfee, *Price Discrimination*, in 1 ISSUES IN COMPETITION LAW AND POLICY 465, 468-69 (2008), available at http://www.mcafee.cc/Papers/PDF/ABAPriceDiscrimination.pdf (the author seems to disagree with Pigou's division, and indicates that "first and third degree price discrimination are each examples of where different groups of consumers are charged different prices for the same good, while second degree price discrimination refers to instances where consumers in a market are presented with the same set of price and quantity options and 'self-select' into different groups." Instead, the author suggests that "[a] more modern and perhaps more useful delineation among the various types of price discrimination designates the old first and third degree of price discrimination as 'direct price discrimination' and the second degree as 'indirect price discrimination'".) (last visited Oct. 16, 2010).

548 Meurer, *supra* note 454, at 850.

549 See McAfee, *supra* note 547, at 465.

From an economic point of view, "discriminatory prices will be required for an optimal allocation of resources in real life situations".[550] As William Baumol and David Bradford observe, "social welfare will be served most effectively not by setting prices equal or even proportionate to marginal costs, but by causing unequal deviations in which items with elastic demands are priced at levels close to their marginal costs."[551] "The reason for the difficulty into which uniform pricing is likely to fall is that real life firms do have at least some monopoly power and can use marketing techniques to keep their markets separate, so that price discrimination is generally possible."[552]

Arguably, price discrimination can exist when three conditions are met. As Michael Meurer points out:

> Effective price discrimination requires satisfaction of three conditions: (1) the seller has market power; (2) the seller can sort customers according to their preferences; and (3) customers cannot arbitrage away price differentials. In the absence of market power a seller would find that competitors would be eager to serve his disfavored customers and any price differential would collapse. The seller must have information on consumer preferences or he will not know who to discriminate against. Arbitrage means that favored customers can buy in order to resell to disfavored customers. The presence of arbitrage undermines price differentials.[553]

Meanwhile, it is admitted that "[i]n most markets price discrimination is either infeasible or feasible only to a limited degree since these requirements are not fully met".[554] In respect of copyright law, the exclusive rights authorized to right holders provide a source of potential market power,[555] which makes price discrimination possible. On the other hand, however, copyright limitations within copyright laws hinder price discrimination. For instance, the first sale doctrine allows the first buyer of a copyrighted work to resell it. This might be

---

550 LOUIS PHILIPS, THE ECONOMICS OF PRICE DISCRIMINATION 1 (NY: Cambridge University Press 1981).

551 William J. Baumol & David F. Bradford, *Optimal Departures from Marginal Cost Pricing*, 60 (3) THE AMERICAN ECONOMIC REVIEW (AER), 265, 267 (1970).

552 PHILIPS, *supra* note 550, at 1.

553 Meurer, *supra* note 454, at 850. See also McAfee, *supra* note 547, at 465 (stating that price discrimination needs three conditions, including "consumers differ in their demands for a given good or service, a firm has market power, and the firm can prevent or limit arbitrage").

554 Meurer, *supra* note 454, at 850.

555 Market power is the ability to price above marginal cost. More details see Franklin M. Fisher, *Detecting Market Power*, in ABA SECTION OF ANTITRUST LAW, ISSUES IN COMPETITION LAW AND POLICY 353-72 (vol.1, US: American Bar Association 2008) ; STEVEN E. LANDSBURG, PRICE THEORY AND APPLICATIONS 355 (7th ed., Thomson South-Western 2008).

harmful to price discrimination if the first user purchases the work at a low price and arbitrages by reselling it at a high price. Additionally, the fair use doctrine which permits copying for private use purposes encourages the sharing of copies among family members and friends and this works against price discrimination.[556]

Recent digital technology, however, promotes price discrimination. By virtue of TPMs, copyright holders can control access to copyrighted works and prevent unauthorized copying and sharing. In addition, "[i]mprovements in computing permit tracking customers much more effectively, and computing prices tailored to the customer."[557] By doing this, price discrimination becomes technically possible and relatively easy to implement. Together with the contract law to permit contractual terms suppressing resale and sharing of copyrighted works by users, the conventional obstacles of adopting price discrimination can be successfully overcome.[558]

Price discrimination is, moreover, the key issue in the debate over the balancing of profit and access in the digital environment.[559] Via the access control and price discrimination enabled by TPMs, users only need to pay certain fees depending on various uses of works. For instance, the iTune store offers songs varying from $ 1.29 for "hot" tracks to $ 0.69 for old songs. It is argued that price elasticity is more reasonable and attractive for both right holders and users.

3.4 Assumption II: The Undue Encroachment of Copyright on Users' "Free Zone"

In comparison, another group of scholars stand for the users' benefits. They argue that the control by TPMs on the use of works exceeds copyright. As copyright holders are anxious to place personal and non-commercial use, private copying, distribution, performance, and the display of copyrighted material

556 See Meurer, *supra* note 454, at 850.

557 McAfee, *supra* note 547, at 483.

558 See Meurer, *supra* note 454, at 851 (indicating that "the fortunes of price discriminators may be bolstered by contract law. Resale and sharing of works by users can be suppressed by contract terms.").

559 See *id.* at 849 (indicating that "[a]cademics and policy-makers have debated the proper copyright response to digital technology by assessing and balancing issues of profit and access, but they have largely neglected what should be a key issue in the debate: price discrimination.").

under economic exploitation in the market, the conventional "free use zone" of private use is being turned into a new opportunity for profit.[560]

In addition, the perfect control offered by TPMs forms a "panoptical prison" that enables copyright holders to monitor, record, and restrict what people read, listen to, or watch in the private sphere. This control greatly harms user's fundamental rights, including user's autonomy in the use of works and the right to privacy.[561]

They further argue that copy-preventing control by TPMs undermines the free speech value of private copying.[562] In this section, I intend to review their main arguments against undue copyright expansion in the digital environment.

3.4.1 Control on the Use of Works Exceeding Copyright

Traditionally, the scope of copyrights granted to right holders by legislation has been limited to the marketing of works. Exclusive copyright refers to the exploitation of works in the market place. In comparison, the use of works by end users exerts the enjoyment function of works rather than the monetary exploitation function.[563] Thus, the use of works is conventionally outside of the scope of copyright.

However, as digital technology enables end users to copy works in large sale and share them with each other, "some copyright owners regard personal uses as implicating the same core reproduction right as commercial uses do. Insofar as there is a commercial demand for personal use copies, copyright owners may want compensation."[564]

In response, right holders have sought to expand their statutory rights to control the access to works and prevent any unauthorized copying. Armed with TPMs and the right to prevent circumvention, copyright holders can actually monitor and measure users' consumption and can argue for a pay-per-use privilege to maximize the economic exploitation of their copyrighted works. "Copyright is now seen as a tool for copyright owners to use to extract all the potential commercial value from works of authorship, even if that means that

560 See Hamilton, *supra* note 178, at 615.

561 See Julie E. Cohen, *DRM and Privacy*, 18 BERKELEY TECH. L. J. 575 (2003) (stating the importance of intellectual privacy, and the ability to use the works without surveillance is a crucial aspect of liberty).

562 See Tushnet, *supra* note 45.

563 See *supra* section I.3 of Introduction.

564 Pamela Samuelson et al., *supra* note 171, at 54.

uses that have long been deemed legal are now brought within the copyright owner's control."[565]

In respect of the control on the use of works, some commentators argue that the measures have already exceeded copyright.[566] As Neil Netanel observes, "the DMCA enables content providers effectively to control access to content, not merely control uses that fall within the parameters of copyright holders' exclusive right".[567] Moreover, "nothing in the Act requires content providers to use that technology only for copyrighted works or only for portions of content that are protected by copyright."[568] Therefore, "[e]ven if copyright's traditional safety valves did continue to afford some semblance of First Amendment protection, those limitations on copyright-holder prerogatives are poised to give way before a technology and contract-based 'paracopyright'."[569]

There have been similar concerns regarding the negative effect of TPMs control on access and the constitutionality of "para-copyright" in judicial cases. For example, in *Stevens v. Kabushiki Kaisha Sony Computer Entertainment*, the Australian High Court considered the disputed TPMs to be unconstitutional "para-copyright" measures which violated protections on freedom of speech.[570]

3.4.2 Perfect Control in the TPMs "Panoptical Prison" Harms User's Fundamental Rights

Conventionally, the defence of user's fundamental rights and freedoms, and freedom of expression and the right to privacy in particular, have constituted one of the main justifications for the adoption of limitations on copyrights. However, the para-copyright provisions of TPMs, especially the perfect control on access to works, have the potential to greatly harm user's fundamental rights.

---

565 LITMAN, *supra* note 176, at 14.

566 About the over-expansion of copyright and the dangers that the expansion poses to users of works, see e.g., H. Laddie, *Copyright: Over-Strength, Over-Regulated, Over-Rated?* 5 EIPR 253 (1996); A. Mason, *Developments in the Law of Copyright and Public Access to Information*, EIPR 636 (1997); Jeremy deBeer, *Constitutional Jurisdiction over Para-copyright Laws*, in IN THE PUBLIC INTERESTS: THE FUTURE OF CANADIAN COPYRIGHT LAW 89-124 (Michael Geist ed., Toronto: Irwin Law 2005).

567 See Neil Weinstock Netanel, *Locating Copyright Within the First Amendment Skein*, 54 STAN. L. REV. 1, 75 (2001).

568 See *id.*

569 See *id.* at 22.

570 See *Stevens v. Kabushiki Kaisha Sony Computer Entertainment*, 221 ALR 448 (2005).

3.4.2.1 User's Autonomy in the Use of Works

Generally speaking, individual's right to autonomy refers to the ability to conduct life in a manner of one's own choosing. "Autonomy mandates the *de jure* right and the *de facto* ability to revisit one's situation and the liberty to exit from it or revise it."[571] Personal autonomy can be labeled as self-government, self-determining, or self-directed. In respect of the copyright field, user autonomy indicates the capability of a user to use the copyrighted works on the basis of their own decision guided by their own desire, will, reason, or goal.

This autonomy is a vital content of inherent human dignity. "Personal autonomy is closely interlinked with the broader concepts of human dignity and personal freedom."[572] In *Cossey v. the United Kingdom*, judge Martens stated "[h]uman dignity and human freedom imply that a man should be free to shape himself and his fate in the way that he deems best fits his personality."[573]

Moreover, personal autonomy has been regarded as a general principle of law. For instance, Article 8 of the European Convention on Human Rights (ECHR) indicates that "everyone has the right to respect for his private and family life, his home and his correspondence". As Judge Van Dijk commented:

> The right to self-determination has not been separately and expressly included in the Convention, but is at the basis of several of the rights laid down therein, especially the right to liberty under Article 5 and the right to respect for private life under Article 8. Moreover, it is a vital element of the "inherent dignity" which, according to the Preamble to the Universal Declaration of Human Rights, constitutes the foundation of freedom, justice and peace in the world.[574]

Thus, it is the fundamental rights of a user to possess the freedom to decide when and where to use the work in what manner, as long as this freedom does not violate the law or threaten the rights and freedoms of others.

Notably, a broader concept of personal autonomy is suggested by "left-liberalism". "In addition to the ability and the right to judge for themselves the

---

571 Marina Oshana, Personal Autonomy in Society 72 (England: Ashgate Publishing Limited 2006).

572 N. R. Koffeman, *(The Right to) Personal Autonomy in the Case Law of the European Court of Human Rights*, p.5, available at http://www.staatscommissiegrondwet.nl/userfiles/files/(The%20right%20to)%20personal%20autonomy%20in%20the%20case%20law%20of%20the%20EC%20of%20Human%20Rights.pdf (last visited Oct. 26, 2010).

573 *Cossey v. the United Kingdom*, appl.no.10843/84, ECtHR judgment of Sep. 27, 1990, dissenting opinion of judge Martens, para.2.7.

574 *Sheffield and Horsham v. the United Kingdom*, appl.no.22985/93 a.o., ECtHR judgment of July 30, 1998, dissenting opinion of Judge Van Dijk, para.5.

merits of various ideas of the good life, perhaps people must be encouraged to assume an active role in influencing and fashioning the institutions – economic, education, legislative, social welfare, media, and so forth – that shape their lives."[575]

Analogously, in developing its case law, the European Court of Human Rights (ECtHR) has ruled that "private life" covers a person's physical and psychological integrity, and thus includes multiple aspects of the physical, social and ethnic identity of a person.[576] Furthermore, "'private life' is a broad term encompassing the sphere of personal autonomy within which everyone can freely pursue the development and fulfillment of his or her personality and to establish and develop relationships with other persons and the outside world."[577]

However, as digital technology makes it possible for copyright holders to monitor, record, and restrict what people read, review, look at or listen to online,[578] this crucial portion of human dignity and personal freedom has been greatly threatened. Perfect control by TPMs has raised a challenging question about user autonomy: who should control what and when? When copyright holders become the decision-maker, and unduly constrain or restrict the freedom of using copyrighted works, user autonomy may be incidentally diminished or violated.

3.4.2.2 User's Right to Privacy

Privacy refers to "the desire by each of us for physical space where we can be free of interruption, intrusion, embarrassment, or accountability and the attempt to control the time and manner of disclosures of personal information about ourselves."[579] In other words, privacy indicates limited access to restricted personal information. The individual has the fundamental right to decide and control when and to what degree information about oneself is communicated to others. The Preamble to the Australian Privacy Charter provides that,

> A free and democratic society requires respect for the autonomy of individuals, and limits on the power of both state and private organizations to intrude on that

---

575 OSHANA, *supra* note 571, at 16.

576 See e.g., *Costello-Roberts v. the United Kingdom*, appl.no.13134/87, ECtHR judgment of Mar. 25, 1993, para.36; *Pretty v. the United Kingdom*, appl.no.2346/02, ECtHR[GC] judgment of April 29, 2002, para.61; *S. and Marper v. the United Kingdom*, appl.nos.30562/04 & 30566/04, ECtHR judgment of Dec. 4, 2008, para.66.

577 *Jehova's Witnesses of Moscow v. Russia*, appl.no.302/02, ECtHR judgment of June 10, 2010, para.117.

578 JESSICA LITMAN, *supra* note 176, at 12.

579 ROBERT ELLIS SMITH, BEN FRANKLIN'S WEB SITE 6 (Sheridan Books 2000).

> autonomy .... Privacy is a key value which underpins human dignity and other key values such as...freedom of speech.... Privacy is a basic human right and the reasonable expectation of every person.[580]

The private domain is an important sphere of privacy. "[M]any, though certainly not all, personal uses happen in the privacy of a home, automobile, or other spaces as to which people have reasonable expectations of privacy. Respect for privacy interests support exempting personal uses from the scope of copyright."[581]

The conflict between copyright protection and the user's fundamental right to privacy has emerged with the development of digital technology. Traditionally, copyright protection has not extended to the private sphere of the individual's use of works. Through authorizing exclusive rights on the one hand, and setting various limitations and exceptions, especially the private use limitation, on the other, the copyright system keeps an internal balance between protection of private interests following from the creation of new works and the public interests in access to and use of such works. This balance may be disrupted, however, by expansive copyright protections such as the adoption of TPMs which may clash with users' privacy rights.[582]

In a TPMs system based on digital content transmission, copyrighted content is wrapped up in an encrypted form and only decrypted when a user is authorized to access and use the content in an authorized manner.[583] TPMs systems can identify the private information of a user by virtue of recognizing the user's device, the content the user accesses, or the code used by the user. For example, given the fact that each computer has a unique "IP address", user devices, location, or accessed information can be identified by checking its IP address.[584]

As TPMs can indeed monitor what individuals read, listen to, or view in the private sphere, a conflict has emerged between the use by copyright holders of TPMs to control access to and use of their works, and the right of users to their

580 THE AUSTRALIAN PRIVACY CHARTER, published by the Australian Privacy Charter Group, Law School, University of New South Wales, Sydney, 1994.

581 Pamela Samuelson et al., *supra* note 171, at 54.

582 See Daniel J. Gervais, *Intellectual Property and Human Rights: Learning to Live Together*, in INTELLECTUAL PROPERTY AND HUMAN RIGHTS 22 (Paul Torremans ed., Netherlands: Kluwer Law International 2008).

583 See Paul B. Schneck, *Persistent Access Control to Prevent Privacy of Digital Information*, 87 PROCEEDINGS OF THE IEEE 1239, 1242 (1999).

584 See Michael A. Froomkin, *The Death of Privacy?* 52 STAN. L. REV. 1461, 1490-91 (2000); Jonathan Weinberg, *Hardware-Based ID, Rights Management, and Trusted Systems*, 52 STAN. L. REV. 1251, 1263-68 (2000).

privacy in using the works.[585] Digital technology enables copyright holders to control access, monitor, track, record, and charge for all uses of copyrighted material in each individual's home or other private sphere and this can greatly threaten the user's right to privacy. Indeed, "[o]ur privacy is shrinking quicker than the polar ice cap; technology is eroding it faster than the legal system can protect it."[586] Without effective efforts against undue encroachment of copyright on users' "free zone", privacy – which protects the inviolate personality, the individual's independence, dignity and integrity[587] – will be lost.

3.4.3 Copy-preventing Control Undermines the Free Speech Value of Private Copying

Freedom of expression refers to the ideas of freely holding opinions, imparting and receiving information.[588] It includes not only the communicator's freedom to impart information to others but also the freedom of the recipient, or the public, to seek and receive information.[589]

Free speech constitutes one of the essential foundations of a democratic society. "[D]emocracy requires more than democratically elected rulers; it requires democratic culture."[590] Meanwhile, democratic culture has been defined as popular participation in culture,[591] because a semiotic democracy is "a society

585 See Estelle Derclaye, *Intellectual Property Rights and Human Rights: Coinciding and Cooperation*, in INTELLECTUAL PROPERTY AND HUMAN RIGHTS 145 (Paul Torremans ed., Netherlands: Kluwer Law International 2008).

586 DAVID H. HOLTZMAN, PRIVACY LOST: HOW TECHNOLOGY IS ENDANGERING YOUR PRIVACY xix (CA: Jossey-Bass 2006).

587 See Edward J. Bloustein, *Privacy as an Aspect of Human Dignity: An Answer to Dean Prosser*, 39 N.Y.U. L. REV. 962, 971 (1964).

588 See Article 19 of the 1948 Universal Declaration of Human Rights: "Everyone has the right to freedom of opinion and expression; this right includes freedom to hold opinions without interference and to seek, receive and impart information and ideas through any media and regardless of frontiers." See also Article 10 of the European Convention on Human Rights (ECHR): "Everyone has the right to freedom of expression. This right shall include freedom to hold opinions and to receive and impart information and ideas without interference by a public authority and regardless of frontiers."

589 See MICHEL VERPEAUX, FREEDOM OF EXPRESSION: IN CONSTITUTIONAL AND INTERNATIONAL CASE LAW 44 (Strasbourg: Council of Europe Publishing 2010) (stating that "[f]reedom of communication benefits both the recipients of the message and the sender.").

590 Tushnet, *supra* note 45, at 539.

591 See Jack M. Balkin, *Populism and Progressivism as Constitutional Categories*, 104 YALE L. J. 1935, 1948-49 (1995).

in which all persons are free and able to participate actively, if not equally, in the generation and circulation of meanings and values".[592] Thus, participation by individuals is a vital sector of democracy in society.[593] And a participatory account of democracy requires individuals to understand their own "authorship of the state".[594] Consequently, free speech promotes a "democratic culture",[595] by virtue of guaranteeing individuals' freedom of participation in a democratic society.

Copyright is an engine of free expression. "By establishing a marketable right to the use of one's expression, copyright supplies the economic incentive to create and disseminate ideas."[596] Moreover, the fair use doctrine within copyright law is regarded as the important protective mechanism for free speech.[597] As the case law has, however, increasingly begun to rely on a test of transformative use to determine if a use constitutes fair use,[598] non-transformative private copying which also serves free speech value is becoming equal to piracy.

In fact, "[a]udiences benefit from copied as well as original expressions."[599] The value of a copyrighted work to users "isn't just in its existing somewhere in a bookstore – the value lies in consumers actually hearing or reading it.... Republished work is materially more valuable to readers than the original that they can't get, that costs too much, or that they don't know about...."[600]

Pure copying can create important free speech value which enhances copiers' abilities in the fields of self-expression, persuasion, and affirmation.[601]

---

592 Michael P. Madow, *Private Ownership of Public Image: Popular Culture and Publicity Rights*, 81 CAL. L. REV. 125, 146 (1993).

593 See Daniel P. Tokaji, *First Amendment Equal Protection: On Discretion, Inequality, and Participation*, 101 MICH. L. REV. 2409, 2505 (2003).

594 See Robert Post, *Reconciling Theory and Doctrine in First Amendment Jurisprudence*, 88 CAL. L. REV. 2353, 2367-68 (2000).

595 See Jack M. Balkin, *Digital Speech and Democratic Culture: A Theory of Freedom of Expression for the Information Society*, 79 N.Y.U. L. REV. 1 (2004).

596 See *Harper & Row, Publishers v. Nation Enters.*, 471 U.S. 539, 558 (1985).

597 See e.g., *L.A. Times v. Free Republic*, 54 U.S.P.Q.2d (BNA) 1453, 1472 (C.D. Cal. 2000) (stating that concerns about free speech "are subsumed within the fair use Analysis"); *A & M Records v. Napster*, 114 F. Supp. 2d 896, 922 (N.D. Cal. 2000) ("Free speech concerns are protected by and coextensive with the fair use doctrine.").

598 See Leval, *supra* note 49, at 1111; *Campbell v. Acuff-Rose Music*, 510 U.S. 569, 579 (1994) (adopting the transformative test of fair use).

599 Tushnet, *supra* note 45, at 565.

600 Eugene Volokh, *Freedom of Speech and Intellectual Property: Some Thoughts After Eldred, 44 Liquormart, and Bartnicki*, 40 HOUS. L. REV. 697, 726 (2003).

601 See Tushnet, *supra* note 45, at 568-81.

Copiers often employ other's copyrighted works to express their own thoughts, emotions, or experiences. For example, a child copying a poem in a card to celebrate a friend's birthday; parents adding music into a home video to recall the good old days; a student explaining herself to others by reference to a book or movie which could perfectly describe her own life. Here, agreement serves an important role in the process of enjoyment of works:

> Enjoyment is pleasure through the senses, and what delights the senses is called agreeable ... We should not use the terms what pleases or displeases to express enjoyment and pain...: these terms are too wide, for they can also refer to intellectual pleasure and displeasure, in which case they would not coincide with enjoyment and pain. We can also describe these feelings in terms of the effect that the sensation of our state produces on our mind. What directly (by the senses) prompts me to leave my state (to go out of it) is disagreeable to me – it pains me. What directly prompts me to maintain my state (to remain in it) is agreeable to me – it delights me.[602]

In addition, given some quoted works are too good to be substituted, it is better to copy and quote from them directly. As Wendy Gordon indicates, "[s]ome poems, some ideas, some works of art, become 'part of me' in such a way that if I cannot use them, I feel I am cut off from part of myself."[603] Indeed, sometimes it would be more persuasive to borrow words originating in famous works of literature. This attitude was well embodied in the traditional Chinese culture which regarded copying as a way of showing respect for the wisdom of the past, demonstrating knowledge and judgment, or preserving general access to the national culture heritage.[604]

Moreover, "[a]ll literary works . . . are 'rewritten', if only unconsciously, by the societies which read them; indeed there is no reading of a work which is not also a 're-writing'."[605] In other words, all private uses by individuals are creative. Different readers might have different understandings towards the same work. Even for the same person, he or she might obtain diverse perceptions at different periods.

Therefore, based on the aforementioned observations, the private use limitation, especially private copying, is designed to safeguard the user's freedom of speech and the public's right to information, and to promote the free flow of information. This argument leads to a consequent assumption that copy-

---

602 Paul Guyer, *Anthropology*, VII: 230, quoting ZUCKERT, *supra* note 17, at 239.

603 Wendy J. Gordon, *A Property Right in Self-Expression: Equality and Individualism in the Nature Law of Intellectual Property*, 102 YALE L. J. 1533, 1569 (1993).

604 See ALFORD, *supra* note 291, at 25-29.

605 TERRY EAGLETON, LITERARY THEORY: AN INTRODUCTION 11 (2d ed., US: University of Minnesota 1996).

preventing measures by right holders would harm the important value of free speech.[606] In practice, TPMs and anticircumvention provisions have been criticized for being used to stifle several important public policy priorities as well as legitimate activities rather than stopping alleged copyright infringements. As EFF observes, these measures have been employed to chill free expression and scientific research, jeopardize fair use, and impede competition and innovation as well.[607]

3.5 What is the Place of Private Use in the Digital Copyright Regime?

After reviewing the two aforementioned competing arguments put forth by either copyright holders or users, in this section, I intend to explore the proper place of private use in the digital environment. I will start from the purpose of copyright law, and then evaluate the incentive-based perfect control approach from a dynamic efficiency perspective. Finally, I will illustrate why copyright law should care about private use.

3.5.1 Starting Point: What Is the Copyright Law For?

"Copyright policy, like other major areas of public policy, requires a solid anchoring in fundamental principles. That anchor can only be found through a clear understanding of the purposes of copyright."[608] To clarify the objective of copyright law possesses three functions: it guides policy-making; it assists courts in interpreting copyright law; and it increases understanding and internalization of copyright norms by citizens with proper education.[609]

In order to make a sound decision on the place of private use in the digital environment, the core issue and departure point should be to investigate what the copyright law is for. There is no solid answer in all countries. A divergence between the utilitarian notion in the Anglo-US and natural law in continental

606 See Tushnet, *supra* note 45.

607 Electronic Frontier Foundation (EFF), *Unintended Consequences: Ten Years under the DMCA*, Oct. 2008, available at http://www.eff.org/files/DMCAUnintended10.pdf (last visited Oct. 26, 2010).

608 Daniel Gervais, *A Canadian Copyright Narrative*, in COPYRIGHT LAW: A HANDBOOK OF CONTEMPORARY RESEARCH 49 (Paul Torremans ed., UK: Edward Elgar Publishing Limited 2007).

609 See *id.* at 50.

Europe constitutes two different models of copyright protection. With the harmonization movement of copyright principles at the international and regional levels, it is difficult to draw a distinct and consistent line between the two groups nowadays.[610] This historical divergence has, to some extent, potentially continued to influence the development tendency of modern copyright laws and had an effect on policy making.

Arguably, the Chinese copyright regime adopts a "dual objective". It strives to fullfil the two objectives of providing sufficient protection to copyright holders' interests on the one hand, and promoting the public interests in the transmission, use, and diffusion of knowledge and information on the other.[611] The primary objective of Chinese copyright law is to protect copyright holders' interest in order to provide incentives for them to create. It serves as a means for the end of maintaining the public interest. The latter is actually the ultimate objective which is pursued by the law. Thus, it is not difficult for us to find a clue of incentive-based instrumentalism taken by the Chinese copyright system.

In this section, I will explore the diverse copyright objectives in Anglo-US, *droit d'auteur* countries, and China. My special interest in this issue focuses on how copyright objectives guide policy making and the interpretation of copyright laws. The outcome of the investigation also serves as an underpinning to evaluate the private use issue in the digital environment or decision-making for any other disputes raised by technological development in the future.

3.5.1.1 Historical Divergence of Copyright Laws in Anglo-US and Continental European Countries

Historically, the starting points of copyright protection were apparently distinct in the common law countries and continental European. The former traditionally followed a utilitarian perspective,[612] and prioritized the "social benefit gained from making certain unauthorized uses of copyrighted materials";[613] while the

---

610 See Gillian Davies, *The Convergence of Copyright and Author's Rights – Reality or Chimera?* 26 IIC 964, 964-65 (1995).

611 See Feng Xiaoqing, Zhuzuoquan zhi lifa zongzhi yanjiu [*Research on the Legislative Objective of Copyright*], in Yuedan minshang [CROSS-STRAIT LAW REVIEW], vol.15, Mar. 2007, available at http://www.fengxiaoqingip.com/lunwen/20070725/1499.html (last visited Oct. 6, 2010).

612 About the description of utilitarian theories, see e.g., Frank I. Michelman, *Property, Utility, and Fairness: Comments on the Ethical Foundations of "Just Compensation" Law*, 80 HARV. L. REV. 1165, 1208-09 (1967).

613 See GUIBAULT, *supra* note 120, at 87.

latter followed a naturalist approach,[614] and emphasized on the protection of the author's rights. In this section, I intend to explore the objectives of these two distinct models of copyright laws and their underpinning philosophies.

3.5.1.1.1 Public Interest Oriented Anglo-US Copyright Statutes

The Anglo-US copyright laws focus on the public interest derived from creative works rather than the private benefit of their creators. As Chief Justice Hughes stated in regard to the copyright monopoly granted by Congress, "[t]he sole interest of the United States and the primary object in conferring the monopoly lie in the general benefits derived by the public from the labors of authors."[615]

"Following the utilitarian argument, the primary objective of the copyright system is to promote the social good, by encouraging the creation and the dissemination of new works to the public."[616] This section initially examines the creativity-enhancing objective of the Anglo-US copyright system. I then turn to the underlying philosophy of copyright policy – utilitarianism – and analyze its representation in the legislative process as well as in judicial practice.

3.5.1.1.1.1 The Creativity-enhancing Objectives of Anglo-US Copyright Law

"The common law copyright model is said to be primarily concerned with encouraging the protection of new works." [617] The creativity-enhancing objectives of copyright law are explicitly stated in Anglo-US legislation. The 1709 Statute of Anne was "an Act for the encouragement of learning ...." A similar statement is found in the US Copyright Act of 1790.[618] Furthermore, the requirement of originality is intended to fulfill the "objectives stated in the original Anglo-US legislation of 'encouraging learning', and additionally in the

---

614 See generally, e.g., JOHN FINNIS, NATURAL LAW AND NATURAL RIGHTS (Oxford: Oxford University Press 1980); STEPHEN BUCKLE, NATURAL LAW AND THE THEORY OF PROPERTY: GROTIUS TO HUME (Oxford: Oxford University Press 1991); NATURAL LAW THEORY: CONTEMPORARY ESSAYS (Robert P. George ed., Oxford: Clarendon Press 1992).

615 *Fox Film Corp. v. Doyal*, 286 U.S. 123, 127 (1932).

616 GUIBAULT, *supra* note 120, at 10.

617 ADRIAN STERLING, WORLD COPYRIGHT LAW, para. 16.06 (London: Sweet & Maxwell 1999).

618 The US Copyright Act of 1790 was the first federal copyright statute, approved on May 31, 1790. Its preamble stated that it was "(An Act) for the encouragement of learning...."

US terms of furthering the Constitutional objective of the promotion of science (i.e. knowledge)”.[619]

Besides the statements contained in statutes, the pubic interest oriented objective of copyright protection is also embodied in judicial judgments. In the landmark British case, *Donaldson v. Beckett*, Lord Camden stated his concern about the copyright monopoly, “All our learning will be looked up in the hands of the ...[booksellers], who will set what price upon it their avarice chuses to demand, till the public become as much their slaves, as their own hackney compilers are.”[620] In the landmark US case, *United States v. Paramount Pictures, Inc*, copyright law was regarded as making reward to the owner a secondary consideration.[621] The statement implied that a primary consideration should be given to the public interest through promoting creativity and information dissemination.

3.5.1.1.1.2 The Utilitarian Notion of Copyright

Utilitarianism

The underlying philosophy of the public interest oriented copyright protection approach is utilitarianism. The Anglo-US copyright laws adopt a utilitarian perspective and emphasize the social welfare derived from the creation of works.

Jeremy Bentham, an English moral philosopher and legal reformer, founded the doctrine of utilitarianism. Its main argument is that the highest principle of morality is to maximize utility.[622] According to Bentham, every moral argument must implicitly draw on the idea of maximizing happiness. Even for those absolute, categorical duties or rights, people would have no basis for defending them unless they believe that respecting them would maximize human happiness, at least in the long run.[623]

In the copyright context, utilitarian theory views the grant of exclusive rights to authors as an incentive to create. In other words, “[t]he legal assurance

---

619 John Adams, *Originality in Copyright: A Solution to the Database Problem?* in COPYRIGHT LAW: A HANDBOOK OF CONTEMPORARY RESEARCH 27 (Paul Torremans ed., UK: Edward Elgar Publishing Limited 2007).

620 *Donaldson v. Beckett*, 17 Cobbett's Parl. Hist. 953, 1000 (1813).

621 See *United States v. Paramount Pictures, Inc.*, 334 U.S. 131, 158 (1948).

622 By “utility”, Bentham means whatever produces happiness and prevents suffering. See MICHAEL J. SANDEL, JUSTICE: WHAT'S THE RIGHT THING TO DO? 34 (London: Penguin Books 2009).

623 See *id.* at 34-5.

that exclusive rights in works of the intellect will be conferred on every author is offered as a bait to encourage intellectual productivity."[624]

Utilitarianism in the British Copyright Law

Utilitarianism was strongly represented in the process of constituting the 1709 Statute of Anne. It was entitled "An Act for the Encouragement of Learning, by Vesting the Copies of Printed Books in the Authors or Purchasers of Such Copies, During the Times Therein Mentioned".[625] The statutory copyright was to be shaped by events and forces directed by stakeholders only as a means to an end for themselves rather than for the author. This phenomenon unfolded into two facets. On the one hand, the authorship provided by the Act was primarily promoted by the Stationers as a substitute for a publishing monopoly. On the other hand, the draftsmen of the Act put the authorship in as a weapon against the Stationers' monopoly.

Originally, the Statute of Anne was a result of the Stationers' effort to maintain their monopoly over literary publishing. When the Stationers' endeavor to influence Parliament to renew the Licensing Act[626] failed, they changed their strategy and argued for an Act to protect authorship instead. "Ostensibly their aim was the protection of the creator of the work. In reality, however, they were aware that any property right that might be recognised for authors would eventually be theirs through assignment." [627] Ironically, the constitutors employed the 1709 Statute of Anne to prevent a continuation of the Stationers' monopoly by allowing anyone to become a copyright owner.

Customarily, the 1709 Statute of Anne has been recognized as the first English copyright act and even the first copyright act in the world to protect an author's copyright.[628] An examination of copyright from a historical perspective,

---

624 SENFTLEBEN, *supra* note 34, at 13.

625 1709-10, 8 Anne, Ch. 19.

626 13 & 14 Car. II, c.33. From the Injunctions of 1559 to the final expiration of the Licensing Act of 1662 in 1694, all books printed in England were required to be licensed. As a result, the Stationers' Company possessed monopoly interests over publication becase of the requirement of registration on the Stationers' register was a prerequisite for publication. See LYMAN RAY PATTERSON: COPYRIGHT IN HISTORICAL PERSPECTIVE 47 (Nashville: Vanderbilt University Press 1968); also MAKEEN F. MAKEEN, COPYRIGHT IN A GLOBAL INFORMATION SOCIETY: THE SCOPE OF COPYRIGHT PROTECTION UNDER INTERNATIONAL, US, UK AND FRENCH LAW 5 (Hague: Kluwer Law International 2000).

627 MAKEEN, *id.*

628 An opposing argument is that the Statute of Anne was not the first English copyright act, because the earlier Star Chamber Decrees, the ordinances of censorship during the

however, shows that such an interpretation is wholly untenable for the Act was basically a trade-regulation statute.[629] Ostensibly, the 1709 Statute of Anne was construed as providing for an author's right. Yet, the purpose of the Act was to provide a copyright that would function primarily as a "trade regulation device" – acting in the interest of society by preventing monopoly, and in the interest of publishers by protecting published works from piracy.[630]

Here, "the author was being used as an instrument against the monopolists, to prevent them from having the renewal term."[631] For instance, the statutory copyright was limited to a term of 14 years, and the renewal term was available only to the author in order to deprive the Stationers' continuation monopoly. Moreover, although the Act explicitly provided an author's right for the first time, this was not based on any natural right principle for a perpetual protection, but demonstrated that every one could hold copyright, and not only the Stationers.[632] In addition to that, the statute "wholly ignored the authors of books, and certainly was not intended to confer any additional rights on them".[633]

After failing in Parliament in a continuation of their monopoly, the Stationers turned to the courts by virtue of employing author's rights. They argued that an author had a common-law copyright based on his natural rights. "The recognition of natural rights benefited the publishers because, as the agent of the author, those rights would devolve to the publisher. Moreover, if copyrights derived from these innate rights, they represented property that existed independently of statutory provisions and could be protected at common law in perpetuity."[634]

The intention of the book sellers in maintaining their publishing monopoly through promoting the idea of the author's common-law rights in perpetuity was defeated in a series of strategic legal battles. In the landmark case, *Donaldson v. Beckett*,[635] the judges in the House of Lords were primarily concerned with making the case an effective instrument in destroying the booksellers' monopoly.

Interregnum, and the Licensing Act of 1662 were copyright regulations. It was, however, the first Parliamentary English copyright act and the first copyright act without provisions for censorship. See PATTERSON, *supra* note 626, at 12.

629 *Id.* at 143.

630 See *id.* at 14.

631 *Id.* at 13.

632 *Id.*

633 JOHN FEATHER, PUBLISHING, PIRACY, AND POLITICS: AN HISTORICAL STUDY OF COPYRIGHT IN BRITAIN 64 (NY: Mansell 1994).

634 B. ZORINA KHAN, THE DEMOCRATIZATION OF INVENTION: PATENTS AND COPYRIGHTS IN AMERICAN ECONOMIC DEVELOPMENT 1790-1920, 230 (NY: Cambridge University Press 2005).

635 *Donaldson v. Beckett*, 2 Bro. P.C.129, 1 Eng. Rep. 837 (1774).

"The court ruled that authors held a common-law right in their unpublished works, but on publication that right was extinguished by the statute, whose provisions determined the nature and scope of any copyright claims."[636]

Following the merit of the 1709 Statute of Anne against the monopoly of the book sellers, the idea of a common-law right in perpetuity was rejected by the court. As Lord Chief Justice De Grey pointed out:

> The truth is, the idea of a common-law right in perpetuity was not taken up till after that failure (of the booksellers) in procuring a new statute for an enlargement of the term. If (say the parties concerned) the legislature will not do it for us, we will do it without their assistance; and then we begin to hear of this new doctrine, the common-law right, which, upon the whole, I am of opinion, cannot be supported upon any rules or principles of the common law of this kingdom.[637]

As the publisher's monopoly rights were rejected by the 1709 Statute of Anne and judicial judgments, copyright transferred from stationer's monopoly rights to statutory author's rights. The origin of an author's rights, viewed from the historical perspective of the British copyright act, unfolded through an apparent utilitarian approach. "Ironically, the English system would find its most faithful representation in the United States, especially after Britain rarified the 1886 Berne Convention and amended its laws in line with copyright practices in Continental Europe."[638] This utilitarian notion was received into and further strengthened by the American copyright law.

Utilitarianism in the US Copyright Law

American copyright policies and patterns are utilitarian and pragmatic. Article I, § 8, of the US Constitution explicitly states: "To Promote the Progress of Science and useful Arts, by securing for limited Times to Authors and Inventors the exclusive Right to their respective Writings and Discoveries." [639] Consequently, the monopoly authorization of copyright to authors is not

---

636 KHAN, *supra* note 634, at 231.

637 *Donaldson v. Beckett*, 17 Cobbett's Parl. Hist. 953, 992 (1813).

638 KHAN, *supra* note 634, at 234. One distinct harmonization between the British copyright system and the civil law system is the recognition of author's moral rights in the 1988 Copyright Act. Moral rights were introducted to give effect to Article 6 *bis* of the Berne Convention, which requires the member states to protect authors' right of attribution and integrity. For a discussion of the moral rights in common law countries, except the US, see Ginsburg, *supra* note 74; William R. Cornish, *Moral Rights under the 1988 Act*, 11 (12) EIPR 449 (1989); Robyn Durie, *Moral Rights and the English Business Community*, 2 ENT. L. REV. 40 (1991).

639 Article I, § 8, U.S. Constitution, available at http://www.usconstitution.net/xconst_A1Sec8.html# (last visited Oct. 6, 2010).

primarily intended for their private benefit, but as a means to the end of promoting the public interest in society by providing incentives for authors to create more works.[640]

The utilitarian notion was adopted from the beginning of the framing of American copyright law. In 1783, Connecticut became the first colony to approve an Act for the Encouragement of Literature and Genius. According to its preamble, the Act was agreeable to the principles of natural equity and justice, and it stated that "every author should be secured in receiving the profits that may arise from the sale of his works, and such security may encourage men of learning and genius to publish their writings; which may do honor to their country, and service to mankind."[641] This preamble looks like it favors author's natural rights. As Zorina Khan indicates, however, "the language merely reflected eighteenth-century rhetoric, for the rules and standards were certainly not consistent with the theory of natural rights."[642]

Subsequently, the first federal copyright statute was approved on May 31, 1790. Its preamble stated that it was an Act "for the encouragement of learning, by securing the copies of maps, charts, and books to the authors and proprietors of such copies, during the times therein mentioned."[643] Accordingly, the copyright protection provided by the 1790 Copyright Act served to promote a broader public interest. As one commentator indicates, "[i]t was felt that copyright protection would serve to increase the flow of learning and information: publication would contribute to democratic principles of free speech, and the diffusion of knowledge also would ensure broad-based access to the benefits of social and economic development for both producers and consumers."[644]

Indeed, American copyright policies are carefully calibrated to further the ends of a democracy in society. "Democratic values emphasized equal and widespread access to learning, and the importance of information flows for maintaining political freedom, as well as active participation by all citizens,

---

640 See Jane C. Ginsburg, *A Tale of Two Copyrights: Literary Property in Revolutionary France and America*, 64 TUL. L. REV. 991, 992 (1990) (stating that "the U.S. Constitution's copyright clause, echoing the English Statute of Anne, makes the public's interest equal, if not superior, to the author's. This clause authorizes the establishment of exclusive rights of authors as a means to maximize production of and access to intellectual creations.").

641 COPYRIGHT ENACTMENTS OF THE UNITED STATES: 1783-1906, compiled by Thorvald Solberg, 11 (Washington, D.C.: Library of Congress 1906).

642 KHAN, *supra* note 634, at 234-35.

643 COPYRIGHT ENACTMENTS OF THE UNITED STATES: 1783-1906, *supra* note 641, at 32.

644 KHAN, *supra* note 634, at 235.

whereas strong copyrights impinged on the fullest attainment of these objectives."[645] Given the fact that the ultimate objective of US copyright law is to promote the public good, copyright policy recognizes that the authors' exclusive rights might conflict with those of society to a greater extent, and when that happens, the rights of the public are primary.[646]

Notably, a frequently cited potential conflict in American law is that between copyright protection and the First Amendment of the US Constitution. As David Nimmer points out:

> In some degree [copyright] encroaches upon freedom of speech in that it abridges the right to reproduce the "expression" of others, but this is justified by the greater public good in the copyright encouragement of creative works. In some degree it encroaches upon the author's right to control his work in that it renders his "ideas" per se unprotectible, but this is justified by the greater public need for free access to ideas as part of the democratic dialogue.[647]

When the copyright act conflicts with the First Amendment, the primacy of the First Amendment mandates that the copyright act be deprived of effect. "Rather than strike down an entire act as overbroad in such a situation, the judiciary prefers to interpret such a statute as narrowly as needed to preserve it for the effectuation of those of its purposes deemed consistent with the Constitution."[648]

Finally and most intriguingly, some American commentators has widely condemned moral rights, and criticized the idea that authors should have a natural right to control their work as a dangerous "charter for private censorship".[649] They argue that the addition of moral rights to copyright runs afoul of the First Amendment,[650] because moral rights are inimical to the values of innovation and free speech.[651] Furthermore, the idea of an inborn right of authorship would harm the interests of publishing industries. Therefore, it is no wonder that US had refused to recognize authors' moral rights for a long time

---

645 *Id.* at 223.

646 See *Myers v. Callaghan*, 5 F. 726 (1881).

647 David Nimmer, *Does Copyright Abridge the First Amendment Guarantees of Free Speech and Press?* 17 U. C. L. A. L. REV. 1180, 1192-93 (1970).

648 *Triangle Pubs. v. Knight-Ridder Newspapers*, 445 F. Supp. 875 (1978).

649 See Peter Jaszi, *Toward a Theory of Copyright: The Metamorphoses of "Authorship"*, 42 DUKE L. J. 455, 497 (1991).

650 See Rebecca Tushnet, *Copyright as a Model for Free Speech Law: What Copyright Has in Common with Anti-Pornography Laws, Campaign Finance Reform, and Telecommunications Regulation*, 42 B. C. L. REV. 1 (2000).

651 See Lawrence Adam Beyer, *Intentionalism, Art, and the Suppression of Innovation: Film Colorization and the Philosophy of Moral Rights*, 82 NW. U. L. REV. 1011 (1988).

until joining the Berne Convention in 1988.[652] However, even though US copyright law does recognize personal rights, authors' moral rights still lack adequate protection,[653] and judges are often relegated to searching for some rough analogues under the laws of defamation, privacy, or contract.[654]

3.5.1.1.2 Author-oriented Continental European Copyright Laws

In contrast to the utilitarian perspective in the Anglo-US copyright laws, an individualistic conception of the naturalist approach is adopted in *droit d'auteur* countries. Among all Continental European countries, France is the most representative example of this model.

The French copyright system is well known as the opposite of that in the US. "In the French tradition, Parliament has repudiated the utilitarian concept of protecting works of authorship in order to stimulate literary and artistic activity."[655] In French copyright law, the author is protected in his status as a creator. A sacred bond between authors and their works constitutes the source of a strong literary property right. Authors' rights in their works are regarded as not only "property by appropriation, but property by nature, by essence, by entirety, by the indivisibility of the object from the subject".[656]

This section first explores the author-centric objectives of copyright law in *droit d'autheur* countries. Next, it examines the natural rights rationale mainly based on the philosophies of Locke's labor theory and Kant's personality theory.

3.5.1.1.2.1 The Author-centric Objectives of Copyright Law in *droit d'auteur* Countries

Products of the French Revolution, the French copyright laws of 1791 and 1793 were said to enshrine authors' exclusive rights which flow from their preferred

---

652 For a discussion of the obstacles of incorporating moral rights into the US copyright regime, see Roberta Rosenthal Kwall, *Copyright and the Moral Right: Is an American Marriage Possible?* 38 VAND. L. REV. 1 (1985).

653 See ROBERTA ROSENTHAL KWALL, THE SOUL OF CREATIVITY: FORGING A MORAL RIGHTS LAW FOR THE UNITED STATES xiii (California: Stanford University Press 2010) (stating that American copyright law rewards economic incentives almost exclusively and lacks adequate moral rights protections).

654 See Neil Weinstock Netanel, *Alienability Restrictions and the Enhancement of Author Autonomy in United States and Continental Copyright Law*, 12 CARDOZO ARTS & ENT. L. J. 1 (1994).

655 HENRI DESBOIS, LE DROIT D'AUTEUR EN FRANCE 538 (3d ed. 1978), quoted in Ginsburg, *supra* note 640, n.7.

656 Portalis, Speech to Chamber of Peers (May 25, 1839), quoted in Ginsburg, *id.* n.6.

status as creators.[657] The French Revolutionary decrees of 1791 and 1793 replaced publishers' privileges with authors' statutory literary property rights.[658] "[O]ne of the 'fundamental ideas' of the revolutionary copyright laws is the principle that 'an exclusive right is conferred on authors because their property is the most justified since it flows from their intellectual creation.'"[659] In essence, "the very nature of this property, entirely personal, entirely moral, entirely united with the creator's thought."[660]

Subsequent French copyright laws developed personalist doctrine to recognize authors' moral rights, including the right of disclosure, the right of retraction, the right of attribution, and the right of integrity. These "moral rights" are, at least in theory, perpetual and inalienable. For instance, these personalist provisions have made the text of the 1957 French law "sounds like the proclamation of the rights of men. These are obviously proclaimed to be natural rights independent of statute. . . ."[661]

Emerging at the end of the 19th and the beginning of the 20th centuries, these moral rights doctrines provoked theoretical and practical divergences between the French and US copyright regimes.[662] Self-interested author-centric monopoly privileges had emerged not only as the keystone of the "French system of literary property", but they also shaped most of the copyright statutes of the civil law countries as well as international copyright laws in the twenty-first century.[663]

The author-centric objectives of copyright law continued to be used after the forming of European Union. For example, in *RTE v. Commission*, the Court of First Instance made a remarkably bold statement that the "essential function [of copyright] … is to protect the moral rights in the work and ensure a reward for the creative effort [of the author]".[664] The statement was regarded as the first

---

657 See Ginsburg, *id.* at 991.

658 See MAKEEN, *supra* note 626, at 9-10 ("Although the 1791 and 1793 laws recognised only two exclusive rights, namely public performance and reproduction, the French courts gave them a wide interpretation to protect literary, dramatic, musical and artistic works against all types of exploitation, and accordingly they were able to survive in that form for over one and a half centuries.").

659 Ginsburg, *supra* note 640, at 992.

660 Alphonse de Lamartine, *On Literary Property*, Report to the Chamber of Deputies (1841), in 8 OEUVRES COMPLÈTES 394, 405 (1842), quoted in Ginsburg, *supra* note 640, n.6.

661 Rudolf Monta, *The Concept of "Copyright" Versus the "Droit d'Auteur"*, 32 S. CAL. L. REV. 177, 178 (1959).

662 See Ginsburg, *supra* note 640, at 996, n.19.

663 See KHAN, *supra* note 634, at 229; See MAKEEN, *supra* note 626, at 9.

664 See para.71 of the judgment in Case T-69/89 *RTE v Commission* [1991] ECR II-485.

time the Community judicature had said anything about the real purpose of copyright and its underlying justification.[665] The argument regarding the "essential function" of copyright was cited approvingly by the Court of Justice, when it confirmed the judgment of the Court of First Instance.[666]

Incidentally, in continental European countries, a higher standard of creativity is required.[667] For instance, French law provides originality and eligibility for a work to be protected if the work bears the imprint of its author's personality. Similarly, German law requires a work to be a "personal intellectual creation".[668]

3.5.1.1.2.2 The Natural Rights Rationale of Copyright Protection

The author-centric copyright protection approach in Continental European countries is based on the principle that "the most sacred, the most unassailable and the most personal of possessions is the fruit of a writer's thought".[669] The foundation for such a system is built upon an understanding of the philosophical justifications for property rights to literary creations. The leading philosophical justifications of author's natural rights in civil law countries are mainly based on labor theory and personality theory.

The natural law approach to copyright follows the lines of Locke's elaboration of the justification of physical property in his *Second Treatise of Civil Government* of 1690:

> Whatsoever then he removes out of the state that nature hath provided, and left it in, he hath mixed his labour with, and joined to it something that is his own, and thereby makes it his property. It being by him removed from the common state nature hath placed it in, it hath by this labour something annexed to it, that excludes the common right of other man: for this labour being the unquestionable property of the labourer, no man but he can have a right to what that is once joined to, at least where there is enough, and as good, left in common for others.[670]

---

665 See DAVID T. KEELING, INTELLECTUAL PROPERTY RIGHTS IN EU LAW: FREE MOVEMENT AND COMPETITION LAW 268 (Vol. I, NY: Oxford University Press 2003).

666 See para.28 of the Joined Cases C-241/91 P and C-242/91 P *RTE and ITP v Commission* [1995] ECR I-743.

667 See KEELING, *supra* note 665, at 269.

668 DAVID I. BAINBRIDGE, INTELLECTUAL PROPERTY 43 (4th ed., London: Pitman Publishing 1999).

669 Ginsburg, *supra* note 640, at 996.

670 JOHN LOCKE, SECOND TREATISE OF CIVIL GOVERNMENT, Chap. 5, sec. 27, 1690 (Raleigh: Hayes Barton Press 2006).

Although Locke's argument originally referred to physical property, his labor theory has been widely applied to justify intellectual property.[671] As the author devotes his intellectual labor to the creation, his work becomes his property. The natural law approach to copyright holds that "all human beings who create works of the mind are entitled to a specific right embracing protection of their moral and economic interests and covering all use of their works".[672]

In Locke's view, the labor is "pains",[673] namely "something unpleasant enough so that people do it only in the expectation of benefits".[674] By virtue of defining labor as an unpleasant activity, the normative proposition of the labor theory can be said to be separate from the instrumental argument. Pursuant to the proposition of the labor theory, the unpleasantness of labor should be rewarded with property. Consequently, the rewarding of labor is a moral or ethical imperative, which is not based on any consideration of the effects of creating property rights. In contrast, the instrumental argument is directly concerned with those effects. It rewards the unpleasantness of labor with property as a means of motivating people to perform their labor to serve the end.[675]

In addition to the justification of property rights as the fruit of the author's labor, an alternative basis for author's natural rights is that creative works are identified as the explicit representation of an author's personality.

This idea has its roots in Kant's personality theory. In his *Groundwork for the Metaphysics of Morals*, Kant launched a devastating critique of utilitarianism, arguing against treating people as a means of maximizing happiness or any other ends. Instead, he argues that morality is about respecting people as ends in themselves.[676] "According to Kant, an author's words are a continuing expression of the author's inner self. An author's right in his work is thus a fundamentally personal right. It is 'not a right in an object...but an innate

671 See e.g. SENFTLEBEN, *supra* note 34, at 10-12; GUIBAULT, *supra* note 120, at 9; PETER DRAHOS, A PHILOSOPHY OF INTELLECTUAL PROPERTY 41-54 (England: Ashgate Publishing Limited 1996) (*cf.* Drahos argues that labor is not the dominant category in the discussion of the origin of property rights, thus it is somewhat misleading to depict Locke as a labor theorist of property, at 48).

672 J.A.L. Sterling, *Creator's Right and the Bridge between Author's Right and Copyright*, 29 IIC 302, 306 (1998).

673 LOCKE, *supra* note 670, sec. 34, at 1690.

674 Lawrence C. Becker, *The Labor Theory of Property Acquisition*, 73 JOURNAL OF PHILOSOPHY 653, 655 (1976).

675 See Justin Hughes, *The Philosophy of Intellectual Property*, 77 GEO. L. J. 287, 303 (1988).

676 SANDEL, *supra* note 622, at 105.

right, inherent in his own person'."[677] A literary work is a narration of the author's thought addressed to the public in a particular form. The author's right to communicate his thought refers to a fundamental aspect of autonomy and freedom that the "one sole original, inborn Right belonging to every man in virtue of his Humanity".[678]

"Given that the author's rights are intimately bound up with the author's personality and autonomy, Kant viewed those rights are inalienable."[679] Kantian personality theory has profoundly influenced the civilian system of copyright, especially in the recognition of author's moral rights. For instance, Article 6 of the French 1957 Copyright Law declares that moral rights are "perpetual, inalienable and imprescriptible", creating a dual system of moral rights and economic rights.[680] Moral rights, including the right of disclosure, the right of paternity, the right of integrity, and so on,[681] enable the author to maintain the "indestructible creational bond" that exists between his personality and the work.[682]

In sum, both utilitarianism and the natural rights rationale possess their limitations. The most glaring criticism of utilitarianism is that it fails to respect individual rights. It uses people as means to an end rather than viewing them as ends in themselves. In comparison, natural rights are criticized as being self-interested and inimical to the values of innovation and free speech.

As the movement of harmonization of copyright principles at the international level, continues, countries from the *droit d'auteur* tradition have begun to emphasize public interest considerations, while countries from the common law tradition are starting to pay attention to author's moral rights

---

677 Neil Weinstock Netanel, *Why Has Copyright Expanded? Analysis and Critique*, in NEW DIRECTIONS IN COPYRIGHT LAWS 29 (Fiona Macmillan ed., Vol.6, UK: Edward Elgar 2007).

678 I. KANT, THE PHILOSOPHY OF LAW: AN EXPOSITION OF THE FUNDAMENTALS OF JURISPRUDENCE AS THE SCIENCE OF RIGHT 56-7, 129-31 (W. Hastie, trans., Edinburgh: T & T Clark 1887), quoting Netanel, *id.* at 30.

679 Netanel, *id.*

680 See CATHERINE COLSTON & KIRSTY MIDDLETON, MODERN INTELLECTUAL PROPERTY LAW 388 (2nd ed., UK: Cavendish Publishing Limited 2005).

681 See Kwall, *supra* note 652, at 5-16 ("The moral right doctrine generally is said to encompass three major components: the right of disclosure, the right of paternity, and the right of integrity. Some formulations of the moral right doctrine also include the right of withdrawal, the right to prevent excessive criticism, and the right to prevent assaults upon one's personality.").

682 See Adolf Dietz, *The Artist's Right of Integrity under Copyright Law: A Comparative Approach*, 25 IIC 177, 182 (1994).

beyond the consideration of mere economic benefits.[683] There is now a growing body of literature that questions the accuracy of the traditional diverse portrayals of copyright laws in common law countries and in civil law *droit d'auteur* systems.[684] Consequently, to some extent, it is difficult to draw a distinct and consistent line between the two systems nowadays.

However, those historical discrepancies in the primary objective and legislative orientation of copyright law continue to influence domestic copyright policies. For instance, during the process of copyright expansion, resistance has been relatively lower in author-centric *droit d'auteur* copyright systems. In this framework, copyright limitations are only exceptions,[685] and often subject to sufficient compensation being paid to right holders.[686] In contrast, every step of copyright expansion in the utilitarian US copyright system has been subject to inquisition based on the Constitution and the First Amendment about the impact on free speech and the public interest. In order to grasp the significance of a legislative orientation or the motives of a certain policy in depth, it is necessary to first understand this historical divergence of copyright laws between the Anglo-US and continental European systems.

3.5.1.2 The "Dual Objective" of the Chinese Copyright Law

Article 1 of the Chinese copyright law clarifies its legislative purpose. This provides not only the sprite of the copyright statute, but also the guiding ideology behind engaging in the task of protecting copyright.[687]

> This Law is enacted, in accordance with the Constitution for the purposes of protecting the copyright of authors in their literary, artistic and scientific works and rights related to copyright, of encouraging the creation and dissemination of works which would contribute to the construction of socialist spiritual and material

---

683 See GUIBAULT, *supra* note 120, at 8.

684 For a discussion of this issue, see generally David Vaver, *The Copyright Mixture in a Mixed Legal System: Fit for Human Consumption?* JURIDICAL REVIEW 101 (2002); Ginsburg, supra note 640; Alain Strowel, *Droit d'Auteur and Copyright: Between History and Nature*, in OF AUTHORS AND ORIGINS: ESSAYS ON COPYRIGHT LAW (Brad Sherman & Alain Strowel eds., Oxford: Clarendon Press 1994) (exploring the historical development of commonalities between droit d'auteur and utilitarian copyright regimes).

685 See SENFTLEBEN, *supra* note 34, at 12.

686 For example, in order to compensate the copyright holders' economic interests effected by private copying, in the 1960s Germany introduced the first copyright levy system in the world, which later became a model for legislators elsewhere.

687 See Shen Rengan et. al. eds., Zhonghua Renmin Gongheguo zhuzuoquan fa jianghua [AN INTRODUCTION TO THE CHINESE COPYRIGHT LAW] 36 (Beijing: Falü chubanshe [Law Press China] 1991).

civilization, and of promoting the development and flourishing of socialist culture and sciences.[688]

Accordingly, the Chinese copyright regime bears a "dual objective" character. It strives to achieve the dual targets of providing sufficient protection to authors' interests while at the same time promoting the public interest in the transmission, use, and diffusion of knowledge and information.[689]

3.5.1.2.1 The Primary Objective of Protecting Author's Interests

Literary and art works are the object of copyright. Rights and interests related to works include copyright, intermediary's rights and public rights (or user's rights). When allocating these rights and interests, Chinese copyright law always takes the author's interest as a first consideration.[690]

The primary objective of Chinese copyright law is to protect the creator's interests. The author occupies a central status within the copyright system based on the following considerations.

First, the object of copyright – works – are the outcome of an author's intellectual labor. The author is the source of creative works. Without respecting creativity and safeguarding their interests, the creative enthusiasm of authors would be restrained. Consequently, without copyright creativity would tend to be like water without a source or a tree without roots.

An author's copyright, including their moral and economic rights, originates directly from the author's own intellectual creation. Moral rights are the embodiment of the author's personality which reflects his ideas, attitudes, or aesthetic sentiments. Economic rights, on the other hand, are the rights regarding the exploitation of works by the author or assignees.[691]

---

688 Article 1 of the 2010 Chinese Copyright Law.

689 See Feng Xiaoqing, *supra* note 611.

690 Feng Xiaoqing, Woguo zhuzuoquan fa dui zuopin quanyi fenpei de junheng sixiang tanxi [*Research on the Balance of Interest Consideration in the Chinese Copyright Law*], 4 Zhishi chanquan [INTELLECTUAL PROPERTY]16, 16 (1996); See also Jiang Ping et al, Zhonghua Renmin Gongheguo zhuzuoquan fa jiangxi [ANALYSIS OF THE CHINESE COPYRIGHT LAW] (Beijing: Zhongguo guoji guangbo chubanshe [China International Broadcasting Press] 1991) (Stating that viewed from the history of the development of Chinese copyright, it evolved from protection by social custom to administrative protection by the feudal government (Guan-fu), and finally to a statute law to protect creators).

691 See Feng Xiaoqing, Shi lun zuozhe zai zhuzuoquan zhong de falü diwei [*The Legal Status of the Author in the Copyright System*], 4 Zhishi chanquan [INTELLECTUAL PROPERTY] 16, 17 (1995).

Consequently, the main value which underpins the Chinese copyright system is the need to confirm the primary status of the authors' interest in the copyright legal relationship, in order to realize the social function of copyright law in protecting the author and promoting cultural communication.[692]

3.5.1.2.2 Ultimate Objective of Maintaining Public Interest

In addition to its primary objective of providing sufficient protection to creators' interests, Chinese copyright law also serves the more important goal of promoting the public interest by encouraging the creation and dissemination of works. The latter is in fact the ultimate objective of Chinese copyright law.

In order to promote creativity, it is necessary to protect rights derived from intellectual property, even those based on the position of natural rights or justice. However, copyright protection obviously has a more fundamental ultimate purpose. It can not be justified to restrict the freedom of the majority merely based on the fact that someone engages in a certain kind of creativity. Therefore, when designing the legal system, it is unreasonable to directly jump to the natural rights rationale. Rather it is necessary to take efficiency, justice, equity, and other factors into consideration. Consequently, it is not permissible to only emphasize the creators' rights while disregarding others freedom limited by those rights.[693]

The "dual objective" of copyright protection is further confirmed by national policy. In order to improve China's capacity to create, utilize, protect, and administer intellectual property, make China an innovative country, and attain the goal of building a moderately prosperous society in all respects, on June 5, 2008, the State Council of the P.R. China issued the Outline of the National IP Strategy.[694] It indicates the need to "[p]roperly balance the need for copyright protection and the need for information dissemination. We must

---

692 See Wu Handong, Guanyu Zhongguo zhuzuoquan fa guannian de lishi sikao [*A Historical Review of the Chinese Copyright Law Notion*], 3 Fa shang yanjiu [LAW & COMMERCIAL INDUSTRY RESEARCH] 44, 49 (1995).

693 Yoshiyuki Tamura, Zhishi chanquan fa de lilun [*The Theory of Intellectual Property Law*], in Zhishi chanquan niankan [INTELLECTUAL PROPERTY RIGHTS ANNUAL JOURNAL] (initial issue, Li Daodao trans., Beijing: Beijing daxue chubanshe [Peking University Press] 2005).

694 The Strategy focuses on five aspects: improving the IP regime, promoting the creation and utilization of IP, strengthening the protection of IPRs, preventing abuses of IPRs, and fostering a culture of IPRs. See the Outline of the National IP Strategy, available at http://www.gov.cn/english/2008-06/21/content_1023471.htm (last visited Oct. 6, 2010).

protect copyrights in accordance with law while at the same time promoting the dissemination of information."[695]

3.5.1.2.3 The Incentive-based Instrumental Theory of Copyright Protection

In spite of the Chinese copyright law having absorbed the legislative approach of the *droit d'auteur* countries and placed higher emphasis on authors' rights than the utilitarian US does, there is little difference between the Chinese and American public interest policies. The instrumental function of copyright is undoubtedly apparent.[696] Given that the ultimate objective of Chinese copyright law is to advance the development and progress of society, the protection of copyright holders' interests serves as a means for this end.[697]

The "dual objective" character of Chinese copyright law is based on a "bilateral standard", namely the state-individual standards of a legal system. This is something of a combination of the "group standard" in Chinese history and the "individual standard" of Western countries.[698] This legal concept focuses on unification and harmonization between national interests and citizen's benefits. It safeguards individual's rights and freedoms on the one hand, and maintains the collective benefits of the state and society on the other.

The modern Chinese copyright system has discarded the traditional "family standard", or "state standard". At the same time, however, it has not rigidly adopted the absolute and indulgent "individual standard". Rather, it has sought to achieve a delicate balance between copyright protection and limitation. By virtue of diminishing conflict between the interests of creators, intermediaries, and users, the Chinese copyright law strives to keep a balance on the basis of safeguarding authors' benefits.[699] In order to promote the development and

---

695 Article 28 of the Outline of the National IP Strategy.

696 See Wang Taiping & Li Changhao, Zhuzuoquan baohu de shuangchong xianzhi –yi sihai gujuan an yin zhi wei jian [*Dual Restriction of Copyright Protection: A Critical Evaluation of the Dead Sea Ancient Literature Case*], 4 Zhishi chanquan [INTELLECTUAL PROPERTY] 58, 61 (2007).

697 Zhou Yong, Shuzi banquan de shidu baohu jiqi shixian [*Appropriate Protection of Digital Copyright and Its Implementation*], in Wenhua chanye yanjiu zhanlüe yu duice [RESEARCH ON CULTURAL INDUSTRY: STRATEGY AND MEASURE] 26 (Xiong Chengyu et al. eds., Beijing: Qinghua daxue chubanshe [Tsinghua University Press] 2006).

698 See Wu Shuchen, falü wenhua yu Zhongguo chuantong falü wenhua lungang [*Outline of Legal Culture and the Chinese Traditional Legal Culture*], in Ruxue yu falü wenhua [CONFUCIANISM AND LEGAL CULTURE] (Zhongguo ruxue yu falü wenhua yanjiuhui [China Confucianism and Legal Culture Seminar] ed., Shanghai: Fudan daxue chubanshe [Fudan University Press] 1992).

699 See Feng Xiaoqing, *supra* note 690, at 16.

prosperity of socialist culture and sciences, it is thus necessary to provide the essential conditions for creativity and incentives for authors, while at the same time ensuring the dissemination of works by intermediaries and access to works by the public.[700]

3.5.2 Evaluation of the Incentive-based Perfect Control Approach

As mentioned above, dealing with the conflict between TPMs control and private use should follow the objective of copyright law. In this section, I intend to employ the copyright objective to evaluate the incentive-based perfect control approach.

Pursuant to the "dual objective" of Chinese copyright law, it is important to recognize the merits of TPMs for copyright protection, while at the same time being cautious of their potential negative effects on creativity from a dynamic efficiency perspective. The ultimate objective of supporting the public interest thus requires Chinese copyright law to maintain an appropriate level of copyright incentives.

3.5.2.1 The Merits of TPMs for Copyright Protection

Copyright law has always evolved in response to technology changes. As society has transitioned to a digital world, the law must address the challenges posed by the Internet and other digital communication technologies. "The popularity of the Internet and the digitalization of information have strained traditional copyright principles and presented many difficult new questions."[701]

More and more individuals are becoming the target of copyright infringement litigation.[702] While users are enjoying the great advantage of digital networks – decentralized and non-mediated information distribution; convenient and prompt reproduction, a transmission process with almost no cost or quality derogation; instant exchange and simultaneous enjoyment by the masses through P2P file sharing – they are also becoming potential online pirates. Innumerable instances of the unlicensed transmission and sharing of

---

700 See Wu Handong, *supra* note 692, at 49.

701 Jordan M. Blanke, *Copyright Law in the Digital Age*, in SOCIAL, ETHICAL AND POLICY IMPLICATIONS OF INFORMATION TECHNOLOGY 224 (Linda L. Brennan & Victoria Johnson eds., US: Information Science Publishing 2004).

702 See Graham, *supra* note 38.

copyrighted works by end users has increasingly constituted a formidable threat to the conventional commercial exploitation of copyright holders.[703]

In response, right holders use an array of TPMs on their digital works, such as passwords or encryption. These self-help protection methods act as "digital locks" which effectively prevent unauthorized access to copyrighted content as well as unauthorized reproduction.

Through access control or copy-prevention TPMs serve as an important means to fight against online piracy. "The use of TPMs reflects a technological attempt by rights holders to prevent the illegal use and copying of their products."[704] Without sufficient protection by TPMs, some copyright holders would hesitate to distribute their works online. This would in the long run decrease the volume of copyrighted material available and be a disservice to consumers.[705]

3.5.2.2 The Limitations of Perfect Control on Creativity Enhancement from a Dynamic Efficiency Perspective

The application of TPMs "could make piracy difficult, but at the same time it could increase the social costs of access to information and knowledge, limit competition, and hinder innovation and creation."[706] As some commentators observe:

> Critics of the TPMs provisions of the DMCA argue that they impede lawful uses of works, such as making a copy of a music CD to listen to on a computer, making a backup copy of a computer program or copying small parts of a DVD movie for the purpose of teaching or criticism. TPMs have also been used to bar the manufacture of competing products, to suppress speech, to limit the first sale doctrine and to fragment markets, such as through regional codes on DVDs. Moreover, the use of TPMs restricts access to works that have already fallen into the public domain.[707]

---

703 See RIAA, *supra* note 39.

704 *Transcript of Hearing on H.R. 107*, in THE DIGITAL MILLENNIUM COPYRIGHT ACT: 2005 SUPPLEMENT 123 (Pike & Fischer 2005).

705 See Ginsburg, *supra* note 43, at 124.

706 Álvaro Díaz, *Free Trade Agreement and Intellectual Property: Impacts and Challenges*, in KNOWLEDGE GENERATION AND PROTECTION: INTELLECTUAL PROPERTY, INNOVATION AND ECONOMIC DEVELOPMENT 130 (Jorge Mario Martínez-Piva ed., NY: Springer 2009).

707 Pedro Roffe, Christoph Spennemann & Johanna von Braun, *Intellectual Property Rights in Free Trade Agreements: Moving Beyond TRIPS Minimum Standards*, in RESEARCH HANDBOOK ON THE PROTECTION OF IP UNDER WTO RULES 302 (Carlos M. Correa ed., UK: Edward Elgar Publishing Limited 2010).

In this section, I intend to explore the limitations of perfect control by TPMs on creativity enhancement from a dynamic efficiency perspective.

3.5.2.2.1 Static and Dynamic Efficiency

The term "efficiency" derives from the Latin word *efficiens*, which means "to obtain something from".[708] There are two different aspects of efficiency in economics – static efficiency and dynamic efficiency.[709]

Static efficiency focuses on minimizing the waste of known or given economic resources.[710] The presumptions include both that resources are given and constant, and that the fundamental economic challenge is to avoid wasting them.[711] Static efficiency, "in terms of profit maximization or resource use efficiency (Pareto efficiency) for the economy, is the holy grail of conventional economics."[712] For example, efficient allocation means "going as far as possible in the satisfaction of wants within resource and technological constraints".[713] Thus, "the static conception of economics reduces the principle of economic efficiency to a simple technical issue of maximization".[714]

In comparison, dynamic efficiency indicates "an inherent quality to adapt, respond or develop new knowledge",[715] or the capacity of an economic system to stimulate creativity.[716] For a long time, the dynamic perspective of efficiency has been almost completely overlooked by traditional economists who have focused on the static aspect of efficiency. However, "dynamic efficiency is the most important aspect of the economic concept of efficiency, especially in the

---

708 See JESÚS HUERTA DE SOTO, THE THEORY OF DYNAMIC EFFICIENCY 2 (Oxon: Routledge 2009)

709 This dichotomy was introduced by Xenophon. See *id.*

710 See *id.* at 4.

711 See MAZZIOTTI, *supra* note 510, at 19-20; see also *id.* at 2 (stating that static efficiency "consists of the sound management of the available (or 'given') resources, to prevent them from being wasted".)

712 DAVID ROONEY, GREG HEARN, THOMAS MANDEVILLE & RICHARD JOSEPH, PUBLIC POLICY IN KNOWLEDGE-BASED ECONOMIES: FOUNDATIONS AND FRAMEWORKS 69 (UK: Edward Elgar Publishing Limited 2003).

713 ROBERT H. SAMET, LONG-RANGE FUTURES RESEARCH: AN APPLICATION OF COMPLEXITY SCIENCE 180 (US: BookSurge Publishing 2008).

714 DE SOTO, *supra* note 708, at 8.

715 WOLFGANG KASPER & MANFRED E. STREIT, INSTITUTIONAL ECONOMICS: SOCIAL ORDER AND PUBLIC POLICY 58 (UK: Edward Elgar Publishing Limited 1998).

716 See DE SOTO, *supra* note 708, at 29.

real world, where equilibrium can never be reached and the ideal of ...static efficiency is by definition unattainable."[717]

The difference between the two conceptual efficiencies lies in the root metaphors they employ. In respect of static efficiency where efficiency is emphasized as a tool to generate stability, predictability, and manageability, the root metaphor is the "balance", while the metaphor "engine" indicates dynamic efficiency when efficiency is emphasized as tool of transformation or growth.[718] In fact, "[t]he essential difference between static and dynamic efficiency is that the first is achieved by working within a given set of initial conditions while the second implies the creation of new initial conditions."[719]

3.5.2.2.2 The Conflict between Static and Dynamic Efficiency

A system operating at maximum short run efficiency may not be dynamically efficient in long term. As Joseph Schumpeter observes,

> A system – any system, economic or other – that at every given point of time fully utilizes its possibilities to the best advantage may yet in the long run be inferior to a system that does so at no given point of time, because the latter's failure to do so may be a condition for the level or speed of long-run performance.[720]

In essence, the conflict of dynamic efficiency versus static efficiency is about creativity, and long term strategic thinking versus short term maximization thinking.[721] I intend to explore the representations of the conflicts between static and dynamic efficiency, including minoritarian bias generated by the interest contest in maximization, and the dynamic inefficiency of perfect control by TPMs.

3.5.2.2.2.1 Interest Contest in Maximization and Minoritarian Bias

Static efficiency is measured in terms of profit maximization or resource use efficiency.[722] The presumption that resources are given and constant in static efficiency inevitably buries the potential interest contest between stakeholders.

---

717 *Id.*

718 See PHILOSOPHY OF TECHNOLOGY AND ENGINEERING SCIENCES 1018-19 (Anthonie Meijers ed., vol. 9, Amsterdam: North Holland (Elsevier) 2009).

719 ROD COOMBS, PAOLO SAVIOTTI & VIVIEN WALSH, ECONOMICS AND TECHNOLOGICAL CHANGE 15 (New Jersey: Rowman & Littlefield 1987).

720 JOSEPH A. SCHUMPETER, CAPITALISM, SOCIALISM AND DEMOCRACY 83 (Unwin Paperbacks 1987).

721 ROONEY, HEARN, MANDEVILLE & JOSEPH, *supra* note 712, at 69.

722 See *id.*

For copyright holders, their rational desire is to maximize their profits, while users wish to maximize their utility.[723]

Potential conflicts are often generated between the tit-for-tat profit-maximizing copyright holders and utility-maximizing users. In accordance with Neil Komesar's theory on minoritarian bias, the concentrated high-stake interests of copyright holders often prevail over the dispersed small-stake interests of users in the legislative process.[724] As a result, copyright holder's exclusive rights have expanded dramatically, while the copyright limitations for the benefit of users have been sharply contracted.

The Premise of Static Efficiency: Given and Constant Resources

Static efficiency analysis operates under the presumption that resources are given and constant. In the copyright context, the resources refer to a sum of social welfare derived from all creative works. The distribution of a fixed resource among various individuals is done under the key assumption that this allocation has no effect upon the total amount to be allocated.[725] Thus, static distribution is only concerned with the benefits of vested interest groups in the short run. "The tension between static and dynamic modes of analysis is particularly important in the case of technological change, in which the initial conditions of the system are always being redefined."[726]

In terms of making decisions about distribution, the two main stakeholders – copyright holders and users – are usually put on the opposite sides of a pipe. The institutional design of the copyright system tends to move backward or forward along a close continuum with the intention of reaching equilibrium with technological development.

---

723 See e.g., David H. Brett, *Digital Knowledge: Copyright Intellectual Property and the Internet*, in WORLD CONGRESS ON INTELLECTUAL CAPITAL READINGS 67 (Nick Bontis ed., Boston: Butterworth-Heinemann 2002) (indicating "profit-maximizing for sellers and utility-maximizing for buyers"); Shai Herzog, Scott Shenker & Deborah Estrin, *Sharing Multicast Costs*, in INTERNET ECONOMICS 173 (Lee W. McKnight & Joseph P. Bailey eds., Cambridge, Mass.: MIT Press 1997) (emphasizing on "the strategic (utility-maximizing) behavior of users, and the profit-maximizing behavior of firms").

724 See KOMESAR, IMPERFECT ALTERNATIVES, *supra* note 51, at 76.

725 See A.B. Atkinson & F. Bourguignon, *Income Distribution and Economics*, in HANDBOOK OF INCOME DISTRIBUTION 41 (A.B. Atkinson & F. Bourguignon eds., vol.1, Amsterdam: North-Holland 2000).

726 COOMBS, SAVIOTTI & WALSH, *supra* note 719, at 15.

For example, the proverbial "three-step test" of the Berne Convention determines the boundary of fair use for the benefit of users in a way that excludes the normal exploitation and legitimate interests of copyright holders.[727] In other words, the normal exploitation and subsequent legitimate interests belong to the copyright holder's exclusive region; while the users' proportion of the creative welfare only exists in certain special cases, provided that this allocation does not unreasonably prejudice the copyright owner' distributive portion.

The conventional copyright limitations attempt to reach a certain equilibrium within this close continuum. For example, the limited duration of copyright protection ensures that the expired works fall into the public domain; the idea/expression dichotomy safeguards the monopoly right of right holders by only referring to the original expressions of works, while leaving the underlying ideas as well as facts free; and, moreover, the copyright limitations, especially the authorization-free and remuneration-free exemptions carve certain uses out of the scope of the copyright holder's exclusive rights.

If these balance of interest mechanisms function well in practice, the expectant equilibrium status can be successfully reached. However, the static distribution in a close continuum can hardly manage a balance in the real world where the two parties are competing for their benefit maximization. The outcomes of the interest contest often end up in a winner-take-all situation, rather than an ideal equilibrium.

This is the case with what is happening in the new round of competition between copyright holders and users in the digital environment. As the copyright is expanding dramatically, the user's "free zone" is quickly shrinking and facing the threat of being overridden by TPMs and contract. In the next section, I will employ the maximization phenomenon in microeconomic theory to explain the conflict of interests between the competing stakeholders. I will also analyze the danger of the private use exemption based on some normative arguments of an institutional approach to law, notably Neil Komesar's theory on capture hypothesis and minoritarian bias in the decision-making process.

The Interest Contest in Maximization

Economics usually assumes that each economic actor tries to choose the best alternative that the constraints allow. This phenomenon can be described

727 See Article 9 (2) of the Berne Convention: "It shall be a matter for legislation in the countries of the Union to permit the reproduction of such works in certain special cases, provided that such reproduction does not conflict with a normal exploitation of the work and does not unreasonably prejudice the legitimate interests of the author."

mathematically as maximization.[728] In this process, people are often premised to be rational, and rationality requires maximization.[729] The conception of the rational behavior of these "rational utility maximizers" recognizes that people choose alternatives that are well-suited to achieving their ends.

The story of the maximization that now happens in the copyright domain is shockingly akin to the aforementioned assumption. The two main stakeholders – copyright holders and users – are both rational people, thus it is rational for them to strain every nerve to pursue their maximum benefit. Although their targets are the same, namely the copyrighted works, their maximum satisfactions are diverse. For copyright holders, their rational desire is to maximize their profits, while users are utility-maximizing. In other words, copyright holders endeavor to extract all the potential commercial value from their works, and seek to bring even uses that have long been deemed legal under their control.[730] Users on the other hand strive to gain the right of access to cultural goods, to broaden the scope of legitimate use, and to re-conceptualize copyright law as also being "a law of user's right".[731]

Potential conflicts are often generated between the tit-for-tat profit-maximizing of copyright holders and the utility-maximizing of users. "Choosing the best alternative that the constraints allow corresponds to maximizing the utility function subject to the feasibility constraint."[732] Notably, this controversy has been explicitly embodied in the two opposite arguments toward the copyright system: that in favor of extending copyright protection to the utmost and that arguing that copyright has already become overprotected as a result of sharing.

The core factor behind the conflict, as some commentators indicate, is greed.[733] Greed exists on both sides. On the one hand, copyright owners argue for extending the term of copyright, and turn every development of user-friendly technologies into a new opportunity for profit. On the other hand, free-rider

---

728 See ROBERT COOTER & THOMAS ULEN, LAW & ECONOMICS 16 (5th ed., Boston: Pearson Education 2008).

729 See Ingmar Persson, *Rationality and Maximization of Satisfaction*, NOÛS, Vol. 22, Dec. 1988, p.537 (stating that a rational people is committed to the utilitarian goal of the maximal satisfaction of rational desires), available at http://www.jstor.org/stable/2215456?seq=1 (last visited Oct. 16, 2010).

730 See LITMAN, *supra* note 176, at 14.

731 See PATTERSON & LINDBERG, *supra* note 9.

732 COOTER & ULEN, *supra* note 728, at 16.

733 Actually, copyright is in bad odor these days. Ginsburg summarizes how copyright got a bad name for itself in one word –greed. See Jane C. Ginsburg, *How Copyright Got a Bad Name For Itself*, 26 COLUM. J. L. & ARTS 61, 61(2002).

users have revealed an increasing rapacity in acquiring and "sharing" unauthorized copies of music, movie, and other works, which threatens the incentives of copyright holders.

In fact, this conflict is not a new topic, but has been around since the origin of copyright law. The overall history of copyright law is a history of expansion.[734] The process is not simple or straightforward, but rather a complex spiral rise from conflict to equilibrium, then from new conflict to new equilibrium. In this spiral rise process, the scope of the copyright domain has spread dramatically. If we compare the existing copyright framework with its ancestors, it is apparent that "[t]oday's copyright law bears scant resemblance to its original formulation."[735] Compared with their early counterparts, the current copyright regime has expanded dramatically in scope, such as in the areas of extended copyright duration, the capacious bundling of exclusive rights, and the constriction of fair use.[736]

The evolution of the copyright regime has proceeded alongside the development of production, replication, and distribution technologies. "Copyright law has a long tradition of adaptation to new technology."[737] The three main historical stages of copyright expansions have all been the outcomes of accommodation to technological advancements. The invention of the first modern mass communication technology – the printing press – was one of the key factors that eventually led to the first copyright statute in 1709.[738] The introduction of photography,[739] motion pictures, piano rolls,[740] radio, television

734 See e.g., BRAD SHERMAN & LIONEL BENTLY, THE MAKING OF MODERN INTELLECTUAL PROPERTY LAW: THE BRITISH EXPERIENCE 1760-1911 (Cambridge: Cambridge University Press 1999).

735 NEIL W. NETANEL, COPYRIGHT'S PARADOX 55 (NY: Oxford University Press 2008).

736 For more on the copyright expansion embodiments, see e.g., NETANEL, *id.* at 54-80 (Outlining the most troublesome areas of copyright's untoward expansion, i.e., copyright duration, creative appropriation, the constriction of fair use, "paracopyright", personal uses, and new technological means of dissemination).

737 Matt Jackson, *The Digital Millennium Copyright Act of 1998: A Proposed Amendment to Accommodate Free Speech*, 5 COMM. L. & POL'Y 61, 72 (2000).

738 See GEORGE H. PUTNAM, BOOKS AND THEIR MAKERS IN THE MIDDLE AGES 343-44 (NY: Hillary House Publishers 1962); MARSHALL LEAFFER, UNDERSTANDING COPYRIGHT LAW §1.2 (2nd ed., NY: Matthew Bender 1995).

739 About the effect of photography on copyright law, see Kathy Bowrey, *Who's Writing Copyright's History*, 18 EUROPEAN INTELLECTUAL PROPERTY REVIEW 322, 326 (1996).

740 See *White-Smith Music Publishing Co. v. Apollo Co.*, 209 U.S. 1(1908) (discussing whether the use of piano rolls infringed copyright).

and cable, and home videotaping[741] in the nineteenth and twentieth centuries led to tremendous changes in the law.[742] In the middle of twentieth century, the emergence of the third revolution in science and technology characterizing by computers and internet networks has profoundly influenced the whole copyright regime.[743]

The conventional interest pattern under copyright law, namely the delicate balance between copyright holders and users, has been shaken again. New digital technology has posed new challenges for the application of copyright law,[744] and also been a powerful factor in the reformulation of the new interest pattern. Digital technology has given users new ways of using copyrighted works, and the chance for copyright holders to profit from these utilities at the same time. In other words, it provides new abilities both for copyright holders to maximize their benefit and for users to maximize their utility. A new round of maximizing competition is thus spreading in the new circumstances created by the development of digital networks.

At the initial stage, new technologies tend to move in favor of users.[745] Digital technology greatly enriches the utilization of copyrighted works and enables users not merely to consume, but to actively participate. One representative shift is that the type of website has updated from Web 1.0 into Web 2.0. If personal websites, publishing, and read only are the characteristics of Web 1.0, blogging, participation, and read/write are the labels of Web 2.0. Based upon user-generated content (UGC), "Web 2.0" encourages visitors to upload and control the content available on websites.[746] As a result, the roles of consumers have potentially changed from passive end-users into content producers and providers.

---

741 *Sony Corp. of America v. Universal City Studio Inc.* 464 U.S. 417(1984) (discussing whether the use of home videotaping infringed a copyright).

742 See Jackson, *supra* note 737, at 72 (n.62 stating these new technologies ultimately led Congress to create a sweeping revision of the law in 1976). See also Jessica Litman, *Copyright Legislation and Technological Change*, 68 OREGON L. REV. 275, 288-89 (1989) (stating the US Congress added the film as a new type of copyrighted works in the revision of Copyright Act in 1912).

743 See generally, e.g., LITMAN, *supra* note 176; MATTHEW RIMMER, DIGITAL COPYRIGHT AND THE CONSUMER REVOLUTION: HANDS OFF MY IPOD (UK: Edward Elgar 2007).

744 E.g., one of the significant challenges for the application of copyright law is raised by the phenomenon of peer-to-peer file sharing. See *Metro-Goldwyn-Mayer Studios Inc. v. Grokster, Ltd.* 125 S. Ct. 2764 (2005). Suggestions on this new type of using, see, e.g., Netanel, *supra* note 6.

745 See Baron, *supra* note 5, at 1245.

746 See generally Tim O'Reilly, *What is Web 2.0?* available at http://www.oreillynet.com/pub/a/oreilly/tim/news/2005/09/30/what-is-web-20 (last visited Nov. 8, 2010).

Furthermore, the availability of "mashes" – "new compositions created by combining the rhythm tracks of one song and the vocal track of another" – is regarded as "the most interesting and entertaining phenomenon of the MP3 libraries on peer-to-peer system".[747] Youtube and other online video-sharing websites have become the exhibition platform of a large number of amateurish mash-ups videos – "fanvids". Just when users are applauding the new era of creation and sharing brought by digital networks, copyright holders are sabotaging this by employing TPMs to disable statutorily permitted uses by the public.

As Paula Baron observes, while initially digital technology seems to move in favor of users, "[i]n the longer term, however, such developments have been considered to contribute to the trend towards stronger ownership rights."[748] Various forms of TPMs offer copyright holders greater means of protecting their works. As Jessica Litman indicates, digital technology makes it possible for copyright holders to monitor, record, and restrict what people read, review, look at or listen to online.[749] Armed with TPMs and the right to prevent circumvention, copyright holders can actually monitor and measure users' consumption and create a pay-per-use privilege that turns the conventional "free use zone" of users into a new opportunity for profit.

Minoritarian Bias

According to Neil Komesar's theory, concentrated high-stake interests often prevail over dispersed small-stake interests in the decision-making process and this generates minoritarian bias.[750] Indeed, as commentators have acknowledged, historically, "the practice of making copyright law is particularly prone to regulatory capture."[751] As Christina Bohannan observes, copyright "confers concentrated benefits on small groups while imposing diffuse costs on large

---

747 See SIVA VAIDHYANATHAN, THE ANARCHIST IN THE LIBRARY: HOW THE CLASH BETWEEN FREEDOM AND CONTROL IS HACKING THE REAL WORLD AND CRASHING THE SYSTEM 104 (NY: Basic Books 2004).

748 Baron, *supra* note 5, at 1245.

749 See LITMAN, *supra* note 176, at 12.

750 See KOMESAR, IMPERFECT ALTERNATIVES, *supra* note 51, at 76; see also Elhauge, *supra* note 53.

751 See e.g., Christina Bohannan, *Reclaiming Copyright*, 23 CARDOZO ARTS & ENT. L. J. 567 (2006); Robert E. Thomas, *Vanquishing Copyright Pirates and Patent Trolls: The Divergent Evolution of Copyright and Patent Laws*, 43 AM. BUS. L. J. 689 (2006); John Cahir, *The Moral Preference for DRM Ordered Markets in the Digitally Networked Environment*, p.7, available at http://www.copyright.bbk.ac.uk/contents/publications/workshops/theme1/jcahir.pdf (last visited Nov. 12, 2010).

groups."[752] She further notes that "[a]s a result of special-interest capture, the Copyright Act confers overly broad rights to copyright owners at the expense of the public interest in having access to creative works."[753]

In the new round of the maximizing contest in the digital environment, the concentrated high-stake interests of copyright holders once again will undoubtedly prevail over the dispersed small-stake interests of users in the legislative process. The substantial resources and powerful influence of large copyright industries, such as Disney, A&M Records, and MGM Studios "allow them to exert tremendous influence on legislative bodies", [754] and they have been very successful in lobbying for stronger copyright protection in face of the fear of widespread infringements on the Internet. As a result, the DMCA and its analogue schemes reflect a regime which allows copyright holders to maximize their economic interests at the expense of the users' benefit in gaining access to works.[755]

3.5.2.2.2.2 Perfect Control and Dynamic Inefficiency

A system that operates to promote static efficiency may not be dynamically efficient in allocating resources to the generation of new knowledge in the long term.[756] This divergence between static efficiency and dynamic efficiency has been found in the context of copyright protection by TPMs.

Static distribution in a close continuum inevitably buries the potential interest contest between the stakeholders on account of limited distributive resources. Stakeholders are presumed to be rational maximizers who are trying to choose the best alternative to achieve their ends. Pursuant to the capture hypothesis and minoritarian bias in the decision-making process, the interest contest between maximizers often ends up with a winner-take-all outcome.

This has been the story of the interest contest between copyright holders and users in the digital environment. The tit-for-tat profit-maximizing of copyright holders and utility-maximizing of users compete for the benefits of digital technology. However, the concentrated high-stake interests of copyright holders prevail over the dispersed small-stake interests of users in the decision-making process.

---

752 Bohannan, *supra* note 751, at 614.

753 *Id.* at 568.

754 See Thomas, *supra* note 751, at 691-92.

755 For illustrations of how copyright industries capture legislators particularly in relation to the US DMCA, see generally *id.* at 712-21.

756 COOMBS, SAVIOTTI & WALSH, *supra* note 719, at 15.

As a result, copyright holders' exclusive rights have expanded to cover the offering and diffusion of works on the internet.[757] In addition, technological measures have also been legalized and employed for the purpose of copyright protection. Armed with TPMs and the right to prevent circumvention, copyright holders can control access to works and establish a pay-per-use privilege, even with respect to fair use that permitted by copyright law.[758]

As the doctrine of fair use, which is commonly referred as an "equitable rule of reason",[759] is unduly undermined and becoming "fared use",[760] the conventional "free use zone" of users has become a new opportunity for profit. TPMs and anticircumvention regulations form the technical and legal foundation for a regime of pay-per-use and perpetual protection.[761] As a result, the traditional balance between copyright owner rights and user privileges in copyright law has been disrupted. A highly expanded copyright and a severely contracted doctrine of fair use are the result.

However, perfect control in static efficiency and the prevention of waste would generate dynamic inefficiency and harm creativity in the long term by raising the cost of secondary creativity. Perfect excludability of copyrighted works would diminish and delay the disclosure of new creations by stifling cultural progress. In light of the fact that copyrighted information is both an input and an output of the process of intellectual creation, the greater the rights or incentives of the first generation, the greater the costs or the lower the incentives of the second.[762]

---

757 The WCT and the WPPT contain "the right of making available" to regulate the new type of utilizing copyrighted works under the digital environment. Member states can implement this right in various ways at the domestic level.

758 See LESSIG, *supra* note 487, at 160 (indicating that copyright holders are using TPMs to restrict fair use and using the right to prevent circumvention to punish those who attempt to evade the restrictions on fair use that they impose through TPMs).

759 S. REP. No. 94-473, 94th Cong., 1st Sess. 62 (1975); H.R.REP. No. 94-1476, 94th Cong., 2d Sess. 65 (1976).

760 See Bell, *supra* note 489.

761 See Nimmer, *supra* note 490, at 220.

762 See Yochai Benkler, *A Political Economy of the Public Domain: Markets in Information Goods vs. The Marketplace of Ideas*, in EXPANDING THE BOUNDARIES OF INTELLECTUAL PROPERTY: INNOVATION POLICY FOR THE KNOWLEDGE SOCIETY 271 (Rochelle Cooper Dreyfuss, Diane Leenheer Zimmerman & Harry First eds., Oxford University Press 2001).

3.5.2.3 Towards an Appropriate Level of Copyright Incentive

Naftel and Spiwak note that "the realization of static economic efficiency may conflict with the attainment of dynamic economic efficiency; thus, it is realistic to expect that the attainment of dynamic economic efficiency may require the sacrifice of some static economic efficiencies in the short-run."[763] In other words, dynamic efficiency should be given priority in decision making:

> Where static and dynamic economic efficiency conflict as public policy goals, policy makers should access the potential for technological change in the industry subject to their jurisdiction. An industry that manifests potential for rapid technological change and innovation should not be guided by policies focused too narrowly on promoting the best use of society's resources from the standpoint of today's technology and resources availability, i.e., static economic efficiency.... Policy makers...should carefully assess policy choices such that dynamic economic efficiency is given substantial priority in the decision making process.[764]

This observation reminds us that our mind should not be clouded by shortsighted self-interested maximization, despite the fact that this may be rational. Rather the focus should be on obtaining greater long-term benefits. As Thomas Nagel advocates, "there is a reason to do not only what will promote that for which there is presently a reason, but also what will promote something future for which it is expected that there will be a reason."[765]

A counterargument against Nagel's future rational orientation could be a "present tense version" of the utilitarian goal of the maximal satisfaction of rational desires. Although copyright holders might acknowledge that leaving certain free zone to users could indeed possess future value, especially when they become users themselves, they usually would not be reconciled with giving up their immediate interests in the short term. They might defend and justify their maximizing activity in this way:

> If I am acting rationally today, I will try to do what maximizes the satisfaction of my present rational desires. Similarly, if I am acting rationally tomorrow, I will try to do what maximizes the fulfilment of the rational desires I will then have. So, if it is not a necessary condition for one's desires being rational that they include a dominant desire to maximize future satisfaction, I may now rationally strive to do what I will

---

763 MARK NAFTEL & LAWRENCE J. SPIWAK, THE TELECOMS TRADE WAR: THE UNITED STATES, THE EUROPEAN UNION, AND THE WORLD 51 (Oregon: Hart Publishing 2000).

764 WALTER G. BOLTER ET AL., THE TELECOMMUNICATIONS POLICY FOR THE 1980'S: THE TRANSITION TO COMPETITION 360 (Prentice Hall 1984).

765 THOMAS NAGEL, THE POSSIBILITY OF ALTRUISM 36 (Oxford: Clarendon Press 1970).

> later rationally strive to undo, for the thrust of my dominant, rational desires may shift from today to tomorrow.[766]

However, a system of rationality that allows this division against a future rational self would fail "to require the most elementary consistency in conduct over time" – the consistency of conduct.[767] As the legal mechanism for the distribution and accumulation of social welfare derived from the creation of works, the copyright regime should maintain consistency of conduct.

Nevertheless, some inconsistencies remain in the attitude towards the consumption of copyrighted works. Copyright holders prefer that each user of a copyrighted work purchase an original, while users "often find it is convenient and economical to purchase a single original and share it within their social group".[768] Therefore, in respect of the copyright regime, this consistency of conduct is intended to keep the fairness of distribution and harmonize the distinct rationalities as a shift of identities between authors and users.

Furthermore, empirical surveys reveal that people sometimes demand much more to give up something that they have than they would be willing to pay to acquire it.[769] From an economic perspective, "[t]he divergence between buying and selling price is called an endowment effect because the price varies depending on the initial assignment of ownership."[770]

Thus, it is not surprising that people may demand a higher price to "sell" a right than they would pay to "buy" the same right.[771] The same story can be said to describe the price of copyrighted works. As William Landes and Richard Posner observe, "[e]very author is both an earlier author from whom a latter author might want to borrow material and the later author himself. In the former role he desires maximum copyright protection for works he creates, but in the latter he prefers minimum protection for works created earlier by others."[772]

Actually, it might be misleading to allege an absolute conflict between authors and users. "[T]hese two 'poles', in reality, are the two sides of the same coin. *Among the users of today are the authors of tomorrow.*"[773] Interest lost in a certain static distribution could be compensated sufficiently through another distribution in the future.

---

766 See Persson, *supra* note 729, at 550 (stating that a rational people is committed to the utilitarian goal of the maximal satisfaction of rational desires).

767 NAGEL, *supra* note 765, at 40.

768 Meurer, *supra* note 454, at 847.

769 See COOTER & ULEN, *supra* note 728, at 91.

770 *Id.* at 91.

771 See *id.* at 87.

772 LANDES & POSNER, *supra* note 55, at 69.

773 SENFTLEBEN, *supra* note 34, at 41.

This analysis implies that "copyright holders might well find it in their self-interest, ex ante, to limit the scope and duration of copyright protection".[774] In other words, the reduced portion of the copyright holder's economic benefit deprived by the copyright limitations could be regained when they utilize others' copyrighted works as users. This also works the other way around; to the extent that a subsequent author or potential author is free to borrow material from a previous one, the cost of the latter author's expression would be reduced and this would promote the creativity of more works. The abundance of copyrighted works will positively enlarge the social welfare generated from creative works for distribution and increase the earnings of each stakeholder.

Accordingly, if we employ the aforementioned observations, it is easy to notice that copyright holders are given a limited exclusive right over their intellectual creations, in such a way that "they can recover the costs of producing that property and receive a sufficient return to give them the incentives to produce that property. A 'sufficient return', however, is not perfect control, and intellectual property does not, therefore, give authors or inventors perfect control".[775]

As Mark Lemley points out the traditional attempt of intellectual property law to eliminate free-riding is groundless and has harmful consequences. In his view, such an attempt is a consequence of the inappropriate extension of real property rhetoric to intellectual property, but overlooks the fact that the subject matter of intellectual property – information – is a non-rival good which cannot be depleted.[776]

To make the point more explicit, I posit this corollary to suggest that the relationship between copyright holders and users is not, and never has been, static but rather highly dynamic and interrelated.[777] The distributed sum of social welfare derived from the creation of copyrighted works is not fixed, but variable. Therefore, the institutional allocation of rights and limitations within the copyright regime should place more emphasis on long term dynamic efficiency rather than short-sighted static efficiency. In order to do so, an appropriate level of copyright protection rather than a perfect control should be carefully maintained.

---

774 LANDES & POSNER, *supra* note 55, at 69.

775 LAWRENCE LESSIG, THE FUTURE OF IDEAS: THE FATE OF THE COMMONS IN A CONNECTED WORLD 97 (NY: Random House, Inc., 2001).

776 See Lemley, *supra* note 521.

777 *Cf.* Baron, *supra* note 5, at 1253, 1254 (indicating that the relationship between private copyright and public use is highly dynamic and complex. In this article, the public use refers to all uses by non-owners, including transformative uses, public good uses, and some private consumptive uses.).

3.5.3 Why Should Copyright Law Care about Private Use?

The dynamic approach dramatically enriches the economic analysis of copyright law and its balance of interest theorem. According to the aforementioned observation, only dynamic efficiency, which places emphasis on developing new knowledge, is truly in accordance with the creative-enhancing objective of Chinese copyright law. Given the fact that private use also contributes to the creativity process, the ultimate public interest objective should not be overlooked during the process of copyright protection. Moreover, from the technological neutrality aspect, the benefits of digital technology should not be monopolized by right holders. Consumers' reasonable expectations for enjoyment should also be respected.

3.5.3.1 Consideration of the Ultimate Objective of Chinese Copyright Law

In order to promote creativity, Chinese copyright law seeks dual objectives. In addition to the primary objective of providing sufficient protection to creators' interests, the Chinese copyright law also serves the more important ultimate objective of protecting the public interest by encouraging the creation and dissemination of works.[778]

The emphasis on creation suggests that "decisionmakers need to pay more attention to long run dynamic efficiency relative to short run static efficiency".[779] Dynamic efficiency suggests a clear role for the long term strategic direction for the allocation of rights and limitations in the copyright regime, namely, the allocation of part of its resources to the generation of new knowledge. "This is required to ensure that the building blocks are put into place today for tomorrow's evolutionary development."[780]

Accordingly, with respect to the dramatic expansion of copyright in the digital environment, the Chinese copyright regime should be *"For Instrumentalism, Against Proprietarianism"*.[781] Proprietarianism has deeper roots in the natural law approach which often refers to those theories of justice

---

778 See *supra* section 3.5.1.2.

779 See ROONEY, HEARN, MANDEVILLE & JOSEPH, *supra* note 712, at 69.

780 *Id.*

781 Citing the subtitle of chapter 9 – "Intellectual Property: For Instrumentalism, Against Proprietarianism", see DRAHOS, *supra* note 671, at 199.

which make natural rights their centrepiece.[782] The chief characteristic of proprietarianism is that it assigns to property rights a fundamental and entrenched status and a priority ranking over other kinds of rights and interests.[783] Proprietarianism states that the ownership privileges should trump the public interest.[784] Thus, "[p]roprietarian sentiments lie behind the expansion of intellectual property rights."[785] As the result of the influence of proprietarian thinking, the scope of the exclusive rights of right holders in copyright regime has constantly increased.

By contrast, instrumentalism requires a strongly articulated conception of the public purpose and advocates duty-bearing privileges. John Dewey uses the term instrumentalism to refer to pragmatism.[786] Instrumentalism in law refers to the notion that law is a tool to realize a goal. Accordingly, copyright law is thus understood as a means to reach the end of promoting creativity. The instrumentalist approach to copyright law is not primarily concerned with rewarding authors for their own sake, but rather the overall welfare of society constitutes the centre of gravity.

Moreover, the postmodern deconstruction of works and "the death of the author" arguments further add complementary challenges to the proprietarian notion of copyright.[787] Post-modernist literary criticism regards authors as tyrants, who impose their meaning on texts, and impede the free circulation, free manipulation, free composition, decomposition, and recomposition of fiction.[788]

The nature of the work is decentralized, multiple, and unenclosed. It is the work rather than the author which is alive and speaking to the reader. Only when the author is dead, or dethroned, can the reader enjoy rebirth. Readers give meaning to the texts they peruse; reading itself becomes a creative act.

---

782 See PHILIP PETTIT, JUDGING JUSTICE: AN INTRODUCTION TO CONTEMPORARY POLITICAL PHILOSOPHY 75 (London and Boston: Routledge & Kegan Paul 1980).

783 See DRAHOS, *supra* note 671, at 200.

784 See *id.* at 202.

785 *Id*, at 211.

786 See JOHN DEWEY, PHILOSOPHY AND CIVILIZATION 26 (NY: Minton, Balch & Co. 1931); See also M.G. WHITE, THE ORIGIN OF DEWEY'S INSTRUMENTALISM (NY: Morningside Heights 1943).

787 See Ginsburg, *supra* note 8, at 7 (observing that post-modernism has moved from English and Comparative Literature departments to the halls, or at least the journals, of law schools, bringing with it a complementary challenge to copyright).

788 See David Lange, A*t Play in the Fields of the Word: Copyright and Construction of Authorship in the Post-Literate Millennium*, 55 L. & CONTEMP. PROBS. 139, 143 (1992).

Reception becomes regeneration, and the distinction between consumptive and transformative use can be blurred.[789]

The "death of the author" announced in literary theory has produced a syllogism in copyright rhetoric: Copyright is a consequence of the romantic conception of authorship; romantic authorship is dead; copyright is (or should be) dead, too.[790] Some critics take the somewhat less radical position that copyright need not die, but it must be revised to reflect the reality of a pluralistic authorship process.[791]

"Property rights would be morality's servants and not its drivers."[792] A misguided natural rights approach, together with vestiges of nineteenth-century Romanticism, have pushed copyright in the direction of a full property right that is immune from claims of public access.[793] As Peter Drahos contends, in respect of copyright, we should have an instrumental attitude rather than a proprietarian one.[794] In other words, the exclusive rights in copyright should be replaced by privilege. And those privileges are necessarily accompanied by duties that circumscribe the exercise of the privilege.[795]

Therefore, "[a]ny legislative prohibition on circumvention of TPMs must take into account not only copyright holders' rights, but users' rights and the public interests in access to and use of information and knowledge."[796] From an instrumentalist perspective, it is of crucial importance to retain the private use limitation as an essential duty of the exercise of the copyright privilege.

---

789 See Michel Foucault, *What Is an Author*, in TEXTUAL STRATEGIES: PERSPECTIVES IN POST-STRUCTURALIST CRITICISM 141 (Josue Harari ed., Ithaca: Cornell University Press 1979).

790 See, e.g., Lange, *supra* note 788 (arguing that if intellectual property survives, it will no longer lend itself to a romantic construction of authorship for the purposes of suppressing speech).

791 See e.g., Peter Jaszi, *On the Author Effect: Contemporary Copyright and Collective Creativity*, 10 CARDOZO ARTS & ENT. L. J. 293 (1992) (arguing that the notion of romantic authorship on which copyright law is based fails to reflect the reality of contemporary polyvocal writing).

792 DRAHOS, *supra* note 671, at 223.

793 See Neil Weinstock Netanel, *Copyright and a Democratic Civil Society*, 106 YALE L.J. 283, 307, n.97 (1996) (listing but disagreeing with the assertion by several copyright and literature scholars).

794 See DRAHOS, *supra* note 671, at 199.

795 See *id.* at 199-200.

796 See Jane Bailey, *Deflating the Michelin Man: Protecting Users' Rights in the Canadian Copyright Reform Process*, in THE PUBLIC INTEREST: THE FUTURE OF CANADIAN COPYRIGHT LAW 125-66 (Michael Geist ed., Toronto: Irwin Law 2005).

3.5.3.2 "Consumption Is Itself Productive"[797]

"[C]reativity is most frequently defined as the product of the combination of two or more ideas in a unique manner, resulting in the creation of a singular idea that is both novel and appropriate."[798] According to the quality of implementation of a creative idea, or innovation, creativity can be describe as "radical" or "incremental."[799]

Radical creativity is highly novel and typically unrelated to or a departure from the existing practices or intellectual patterns. In contrast, incremental creativity is relatively minor and often based on an existing concept or the developmental extension of existing practices.[800] In other words, incremental creativity means "modifying something that already exists to make it better".[801] "In essence, an incremental idea represents a logical 'next step,' while a radical idea represents a step in a different direction. The fundamental difference between radical and incremental concepts is the extent to which new and diverse knowledge is incorporated and applied in a unique manner."[802]

Notably, copyright law has never limited its protection to works of radical creativity.[803] As Paul Saint-Amour indicates:

> The "original expression" a work must exhibit in order to win copyright is not a transcendent iconoclasm or singularity but something more modest: the absence of verbatim copying and the demonstrable presence of a modicum of creativity.... Copyright does not even stipulate that "original" works must be different from preexisting ones, only that they be the products of independent creative exertion rather than outright copying.[804]

Generally speaking, the creative process involves two means of production: preexisting works and human capital. Human capital is an essential input in the

---

797 I borrow this expression from Koren, see Koren, *supra* note 48, at 1137.

798 Monique Ziebro & Gregory Northcraft, *Connecting the Dots: Network Development, Information Flow, and Creativity in Groups*, in CREATIVITY IN GROUPS 136 (Elizabeth A. Mannix, Margaret A. Neale & Jack A. Goncalo eds., UK: Emerald Group Publishing Limited 2009).

799 See *id.* at 138.

800 See *id.*

801 JEFF DEGRAFF & KATHERINE A. LAWRENCE, CREATIVITY AT WORK: DEVELOPING THE RIGHT PRACTICES TO MAKE INNOVATION HAPPEN 111 (US: John Wiley & Sons, Inc., 2002).

802 Ziebro & Northcraft, *supra* note 798, at 138.

803 See PAUL K. SAINT-AMOUR, THE COPYWRIGHTS: INTELLECTUAL PROPERTY AND THE LITERARY IMAGINATION 7 (NY: Cornell University Press 2003).

804 *Id.*

creative process.[805] It is a combination of the individuals' intelligence, knowledge, abilities, experiences, skills, time, and energy.[806] The important process of cultivating human capital – education – is not only implemented by educational institutions, such as schools, libraries, or other public institutions, but also by virtue of individuals' private use at home. Private study, such as reading, watching, listening, and communication, is the gateway to all knowledge. It fosters creativity-relevant human capital. By doing so, private use provides the underground nutrition for creativity, and consumption itself possesses productive character.

In a word, we need be cautious about the possible negative effect of perfect excludability by TPMs and the high monopoly price for access to further creative works. Efforts should especially be made to prevent the Matthew effect in the creativity field. Its representations are that "the creatively 'rich' get richer, and the creatively 'poor' stay poor",[807] or that the rich in economy have access, while the poor are excluded. The cultural Matthew effect goes against democratic culture and the cultural prosperity of society as a whole.

3.5.3.3 Technology Neutrality and Consumers' Reasonable Expectations

The principle of "technology neutrality" is introduced in the United Nations Convention on the Use of Electronic Communications in International Contracts, also referred to as the Electronic Communications Convention (ECC).[808] "This principle holds that the law should not discriminate between different forms of technology.... The goal of technology neutrality is important from the

805 See Phillipe Aghion, Cecilia García-Peñalosa & Peter Howitt, *Knowledge and Development: A Schumpeterian Approach*, in ECONOMIC GROWTH AND MACROECONOMIC DYNAMICS: RECENT DEVELOPMENTS IN ECONOMIC THEORY 77 (Steve Dowrick, Rohan Pitchford & Stephen J. Turnovsky eds., NY: Cambridge University Press 2004).

806 See MICHAEL ARMSTRONG, A HANDBOOK OF HUMAN RESOURCE MANAGEMENT PRACTICE 33 (10th ed., UK: Kogan Page Limited 2006).

807 See Isaac Getz & Todd Lubart, *Creativity and Economics: Current Perspectives*, in THE ROUTLEDGE COMPANION TO CREATIVITY 210 (Tudor Rickards, Mark A. Runco & Susan Moger eds., Oxon: Routledge 2009).

808 See Article 9 of the ECC.

standpoint of not stifling development of any technology or unfairly favouring one technology over another."[809]

Given the fact that the principle of technology neutrality reflects regulators' desire to achieve both static and dynamic efficiency,[810] it has been widely introduced in regulatory policies. For instance, the EU communications regulatory framework regards technology neutrality as one of the high-level principles which shows that "policy-makers are keenly aware that technological development in the communications space is not a one-player game".[811] In other words, the development of new technologies includes both healthy competition within the market and also the creativity-enhancing attempts to promote new technology innovation.

Similarly, during the process of constituting the DMCA, some commentators expressed their concerns to the US Congress about the negative effect of anticircumvention regulations on technology neutrality of copyright law:

> Although [they] would be codified in Title 17, [the anticircumvention provisions] would not be an ordinary copyright provision; liability under the section would result from conduct separate and independent from any act of copyright infringement or any intent to promote infringement. Thus, enactment of [the anticircumvention provisions] would represent an unprecedented departure into the zone of what might be called paracopyright – an uncharted new domain of legislative provisions designed to strengthen copyright protection by regulating conduct which traditionally has fallen outside the regulatory sphere of intellectual property law.[812]

Their concerns raised the attention of the legislator. In order to "protect the interests of copyright owners in the digital environment, while ensuring that copyright law remains technology neutral", the Committee on Commerce removed the anticircumvention regulations from Title 17 of the United States Code, and established them as free-standing provisions of law.[813]

---

809 AMELIA H. BOSS & WOLFGANG KILIAN, THE UNITED NATIONS CONVENTION ON THE USE OF ELECTRONIC COMMUNICATIONS IN INTERNATIONAL CONTRACTS: AN IN-DEPTH GUIDE AND SOURCEBOOK 139 (Netherlands: Kluwer Law International 2008).

810 See Bruno Basalisco, Andy Reid & Paul Richards, *Interdependent Innovation in Telecommunications: Risk, Standardization and Regulation*, in REGULATION AND THE EVOLUTION OF THE GLOBAL TELECOMMUNICATIONS INDUSTRY 284 (Anastassios Gentzoglanis & Anders Henten eds., UK: Edward Elgar Publishing Limited 2010).

811 See *id.* at 285.

812 H.R.Rep.No.105-551, at pp.24-25, see THE DIGITAL MILLENNIUM COPYRIGHT ACT: TEXT, HISTORY, AND CASELAW 186-87 (US: Pike & Fischer, Inc. 2003).

813 H.R.Rep.No.105-551, at p.25, see *id.* at 187.

The parallel of the technology neutrality principle in the newly digital environment is Network Neutrality, or Net Neutrality for short. "Net neutrality is one of the key design principles of the Internet that ensures that the network does not discriminate between types of information or the types of parties involved and that information transmitted through the Internet can be equally accessed by all Internet service providers and users."[814] It is the guiding rule that preserves the free and open Internet and protects users' rights to direct online activities based on their personal motivations. It requires the Internet service and content providers to move data in a nondiscriminatory manner.[815]

In the digital environment, "[t]echnology has fostered private use, and it will continue to expand individuals' capabilities to make private use of copyrighted materials. With each new application of technology, new forms of private use will occur."[816] From the technology neutrality aspect, the benefits of digital technology should not be monopolized by right holders, and consumers' reasonable expectation for enjoyment should also be respected. As Jessica Litman indicates:

> Just as technology spurs evolution in the creation and marketing of works of authorship, it causes parallel evolution in the modes of interaction with those works. We don't want to limit copyright owners to the traditional marketing outlets of bookstore and sheet music sales. Similarly, it makes no sense to limit readers, listeners, and players to pianos or analog cassette tapes.[817]

In respect of the TPMs surveillance on the access and use of digital copyrighted materials, Congressman Rick Boucher argues that US copyright law should restore a balance between copyright protection and technological innovation as well as consumer freedom:

> When consumers who purchase digital media can enjoy a broad range of uses of the media for their own convenience in ways which do not infringe copyrights, everyone wins…This principle has been borne out in history time and time again as new technologies have been developed which make the media products ultimately more valuable to consumers. The challenge for content creators is to embrace these new technologies and to create business models around them, rather than shun them by seeing more and more control over the downstream uses of media.[818]

---

814 ROBIN ANDERSEN & JONATHAN GRAY, BATTLEGROUND THE MEDIA 251 (Westport: Greenwood Press 2008).

815 See *id.* at 289.

816 US CONGRESS, *supra* note 534, at 194.

817 Litman, *supra* note 4, at 1911.

818 Rick Boucher, *Copyright and Fair Use: The Present and Future Perspective*, a keynote addressed before the Harvard Journal of Law and Technology Symposium, March 15, 2003.

Therefore, it might be incorrect to arbitrarily argue that all types of unauthorized private use, especially private copying, are copyright infringements. A more cautious and responsible attitude requires comprehensive consideration. Because "technologies that permit copying can either harm or benefit both proprietors and users, depending on several empirically determined factors: the costs of production and distribution, the behavior of producers and consumers in the absence of unauthorized copying, and the effects of copying on subsequent purchasing behavior."[819]

3.6 Summary

Private use is full of uncertainty both in theory and in application. In order to locate a relative clear boundary between copyright exploitation and private use enjoyment, there are at least two vital sub-factors that need to be taken into consideration, including whether the use is commercial or noncommercial, and whether it occurs within the market place or within the private domain. However, as new digital technology provides new abilities both for copyright holders to explore the new market and for users to enjoy new utilities, the conventional boundary between enjoyment and exploitation is becoming blurred in the digital environment.

In response, copyright holders have lobbied legislators to expand their exclusive rights into digital networks, employed TPMs to control access, and prevented users from circumventing them. The perfect control by TPMs has raised fierce debates on the status of private use in the digital environment.

In spite of some positive efforts to harmonize TPMs and private use in the digital environment from both the legislative and judicial perspectives – such as the prerequisite of TPMs protection in the WCT and the WPPT, the saving clauses on TPMs protection in the DMCA, and the restraint application of anticircumvention rules or regarding copyright limitations as users' right in practice – there still remains a lot of controversy over TPMs' control on private use.

Some commentators argue that TPMs control is necessary because the conventional market failure justification for private copying has already lost its importance in the digital environment, and private copying is no longer innocent *de minimas* but large enough in scale to displace sales and constitute competitive exploitation. Moreover, by virtue of TPMs access control and price discrimination, users only need to pay certain fees depending on various uses of

819 US CONGRESS, *supra* note 534, at 198.

works. This new business model with price elasticity is more reasonable and attractive to both right holders and users.

Some commentators, however, argue that TPMs control on the use of works already exceeds copyright. This unconstitutional para-copyright measure greatly harms user's fundamental rights including user's autonomy in the use of works and user's right to privacy. Moreover, given the fact that pure copying could enhance copiers' abilities in areas such as self-expression, persuasion, and affirmation, the copy-preventing TPMs undermine the important free speech value of private copying.

In my point of view, dealing with the conflict between TPMs control and private use – or decision-making for any other disputes concerning private use raised by future technology development – should all follow the objective of copyright law. Chinese copyright law pursues dual targets, regarding copyright protection as the primary objective which serves as a means for the ultimate objective of maintaining the public interest.

Pursuant to the "dual objective" of Chinese copyright law, on the one hand, it is necessary to acknowledge the merits of TPMs for copyright protection, while at the same time being cautious of their limitations on subsequent creativity from a long term dynamic efficiency perspective. Instead, an appropriate level of copyright incentive rather than a perfect control should be maintained.

On the other hand, the ultimate objective of the Chinese copyright law requires the policy maker to pay more attention to long run dynamic efficiency which emphasizes the development of new knowledge. Given the fact that private use is the gateway to all knowledge and fosters creativity-relevant human capital, it also contributes to the creativity process. Meanwhile, the principle of "technology neutrality" reminds us that the benefits of digital technology should not be monopolized by right holders, but rather users' reasonable expectation for enjoyment should also be considered.

Chapter IV Alternative Business Models of Digital Content Provision and Their Social Conditions

Arguably, TPMs are not the sole option for digital content providers to distribute their copyrighted information on the Internet. Technological development has provided many possibilities. In this chapter, I intend to compare the "TPMs & contract" model with the other three alternatives of digital content provision – the "levy system" model, the "device-centric" model, and the "cloud" model. I will illustrate their pros and cons concerning private use, and the social conditions each requires to be met in order for it to be effective. In the last section, I will examine the potential problems in the Chinese practice in seeking to reach these social conditions.

4.1 Four Alternative Business Models

Currently, there are at least four alternative business models of digital content provision – the "TPMs & contract" model, the "levy system" model, the "device-centric" model, and the "cloud" model. Admittedly this is not an exclusive list. There are also other business models, such as Creative Commons' CC license, and advertisement-based open-source businesses. Moreover, with the development of technological and economic conditions, new business models will undoubtedly come into being in the future.

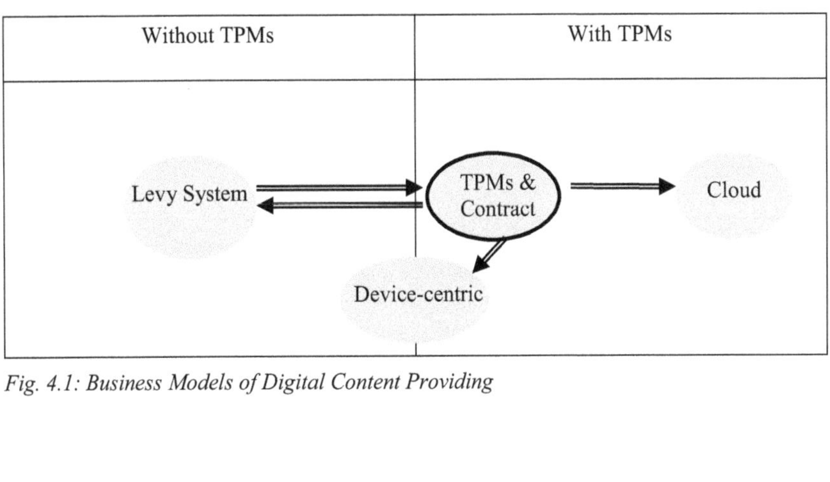

Fig. 4.1: Business Models of Digital Content Providing

In order to make the issues more understandable, in this section, I try to provide a general description and some examples for each business model. In light of the fact that the "TPMs & contract" model has been discussed quite sufficiently in the previous chapters, my focus will here lean more to other three alternatives.

4.1.1 The "TPMs & Contract" Model

At present, the "TPMs & contract" model is the most popular or mainstream approach adopted by copyright holders for the distribution of their digital copyrighted materials. Employing TPMs access control, right holders set contractual terms for the use of their works. Only those who accept the contractual terms get the authorization to use the works. Furthermore, copyright holders can control the ways in which the works are used through TPMs such as through the use of copy-preventing technology to prevent replication.[820]

Usually, whenever you intend to download and install software on your personal computer, there will be a click-wrap license agreement presented on your screen. The agreement includes standard form contract terms regulating use, use restrictions, ownership, liability, reverse engineering and so on. "These electronic standard form contracts are written by the right owners on the model of the 'shrink-wrap' licence and presented to the user on a take-it-or leave-it basis."[821] Unless you accept those terms, you can not continue the installation.[822] After installing the software, you must follow the agreement to undertake the limitations of use by TPMs. In this process, the access control and use control by TPMs and the end user license agreement are perfectly combined.

On account of successful lobbying by the copyright industry, this business model, especially the adoption of TPMs is legitimate and protected by circumvention regulations in copyright law.[823]

---

820 See MACQUEEN ET AL, *supra* note 458, at 230 ("TPMs…can prevent access to and use of a work unless and until such contractual conditions as the producer imposes are met by the would-be user".).

821 GUIBAULT, *supra* note 120, at 204.

822 For example, the end user license agreement of the EndNote X2 indicates that "you must select 'I accept the license agreement' in order to continue with the installation." See EndNote X2 Getting Started Guide, p.14, available at http://www.endnote.com/support/helpdocs/ENX2_GettingStartedGuide_WinMac.pdf (last visited Nov. 11, 2010).

823 See e.g., 17 U.S.C. §1201; Article 11 of the WCT; Article 18 of the WPPT.

4.1.2 The "Levy System" Model

The "Levy system" model has existed for a long time. As the rise of photocopying and analogue recording equipment enabled individuals to make relatively easy and cheap copies of works, the private copying levy was introduced in the 1960s as a compromise between the exploitation interests of right holders and the convenience of users in their enjoyment of works. Private copying continues to be legally allowed in many jurisdictions in exchange for the payment of a levy on copying media.[824]

In the digital environment, as the rampant unauthorized transmission and P2P file sharing of copyrighted works by end users has increasingly threatened right holders' ability to commercially exploit their works, the "TPMs & contract" model has been introduced to safeguard the interests of content providers. "Unlike levies, DRM makes it possible to compensate rights holders directly for the particular uses made of a work, according to the prevailing market price for that work. Where such individual rights management is available where would appear to remain no need, and no justification, for mandatory levy systems."[825]

However, the persistent existence of the unauthorized distribution, exchange, and P2P file sharing of copyrighted materials justifies the copyright levy systems' assumption that private copying of protected works cannot be controlled. The levy system should therefore be extended into the digital environment.[826] Additionally, in respect of the shortcomings of the "TPMs & contract" model,[827] some commentators suggest that right holders should delete TPMs and adopt the copyright levy system as a means of remuneration.[828]

Notably, most European countries have followed the traditional Germany copyright levy systems.[829] In the digital environment, the extension of a general

---

824 See *supra* Introduction I.4.

825 Lucie Guibault, *The Limit of Balancing Interests Through Copyright Levies*, in ETHICS AND LAW OF INTELLECTUAL PROPERTY: CURRENT PROBLEMS IN POLITICS, SCIENCE AND TECHNOLOGY 232 (Christian Lenk, Nils Hoppe & Roberto Andorno eds., England: Ashgate Publishing Limited 2007).

826 See Daniel J. Gervais, *The Price of Social Norms: Towards a Liability Regime for File-Sharing*, 12 J. INTELL. PROP. L. 39 (2004).

827 See *infra* section 4.2.1.

828 See e.g., FISHER III, *supra* note 239, at 199-258; Netanel, *supra* note 6 (advocating a "Noncommercial Use Levy" imposed on P2P-related services and products for the sake of an unrestricted TPMs-free noncommercial use, including P2P sharing).

829 Only three EU member states – Ireland, Luxembourg and the UK have not implemented a copyright levy system for home taping and reprographic reproduction. See Guibault, *supra* note 825, at 236.

levy regime is usually done through the charging of a fee on the sale of digital equipment, devices, or media, such as PCs and MP3 players, or on a subscription to Internet access. For example, Germany imposes a copyright levy of $ 13 plus 16 % in value added tax per new computer sold in the country.[830] In the US, the 1992 Audio Home Recording Act includes a levy system for charging 2 or 3% of the price of digital audio recording devices or blank storage media.[831]

Table 4.1: Portion of Levies in Germany[832]

audio recording devices (except MP3 players)	€1,28	per device
audio recording media	€0,0614	per hour
MP3 players	€2,56	per player
video recording devices	€9,21	per device
video recording media	€0,087	per hour
CD burners	€6,50-7,00	per burner

Moreover, in order to reconcile the existing levy systems of private copying with a new TPMs framework, Article 5 (2) (b) of the EU InfoSoc Directive provides that when calculating the amount of "fair compensation" to right holders, the "application or non-application of technological measures" should be taken into consideration.

830 See Sam Vaknin, *Germany's Copyright Levy*, available at http://samvak.tripod.com/busiweb37.html (last visited Nov. 8, 2010).

831 See 17 U.S.C. §§ 1003, 1004.

832 Quoting Stefan Bechtold, *The Present and Future of Digital Rights Management: Musings on Emerging Legal Problems*, in DIGITAL RIGHTS MANAGEMENT: TECHNOLOGICAL, ECONOMIC, LEGAL AND POLITICAL ASPECTS 614 (Eberhard Becker et al. eds., Berlin: Springer 2003).

4.1.3 The "Device-Centric" Model

In a "device-centric" model, the whole distribution, storage, and usage of digital copyrighted materials heavily rely on or are constrained by the device. "Some services, like Apple's iTunes and Sony Connect, are device-centric: they exist to drive sales of the companies' music players."[833]

"Digital entertainment – and digital music specifically – is already a device-centric market, as evidenced by the iPod's ability to make iTunes the No.1 digital music store in the world."[834] iTunes was introduced by Apple Inc. on January 9, 2001. It is a type of jukebox software that lets users play and organize digital music and video files to create their own digital library. It can connect to the iTune store to preview, purchase and download a user's favorite music, videos and other content. The iTune player is also an interface to manage, transfer, and synchronize the contents among a series of Apple products, such as the Mac, iPod, iPhone, and iPad.[835]

Through years of exploitation, iTunes has become the world's No.1 online music store. As Steve Knopper indicates:

> iTunes emerged as the biggest online retailer, taking more than 70% of the market and dwarfing later competitors like Napster (whose assets had been purchased by another company, Roxio, and turned into a legal service) and stores from Microsoft, MTV, Sony Corp., and Wal-Mart. By April 2008, the iTune Music Store had sold more than 4 billion songs around the world. It ranked No.1, above Best Buy and Wal-Mart, as the top overall music retailer in the United States....[836]

Notably, the initial agenda of the iTune Store was not to make money off 99-cent digital songs, although they were a nice additional source of revenue, but to use the songs to profit from the sale of expensive iPods.[837] In order to attract consumers, Apple Inc. deleted copy-preventing TPMs on its music, thus users could copy and use the purchased application on up to five devices.[838] However,

---

833 TAD LATHROP, THIS BUSINESS OF GLOBAL MUSIC MARKETING 110 (NY: Billboard Books 2007).

834 Antony Bruno, *Phone Face-Off: Manufacturers and Carriers Maneuver to Control the Mobile Future*, in BILLBOARD, p.30, sep. 15, 2007.

835 See *What Is iTunes*? available at http://www.apple.com/itunes/what-is/ (last visited Mar. 4, 2011).

836 STEVE KNOPPER, APPETITE FOR SELF-DESTRUCTION: THE SPECTACULAR CRASH OF THE RECORD INDUSTRY IN THE DIGITAL AGE 179 (NY: Free Press 2009).

837 *Id.*

838 *Cf.* Amazon MP3 started to sell TPM-free MP3 files before iTune Store. The Amazon MP3 Store FAQ states that: "The MP3 files you purchase from Amazon.com do not contain any software (frequently referred to as 'digital rights management' software) that will restrict your use of the file." Available at http://www.amazon.com/gp/help/

for a long period of time, Apple Inc. still tried to retain certain TPMs to make sure that the iTune music could only be used on Apple products, and its products could only use iTune music.[839] This "device-centric" strategy has thus induced anti-competition complaints about the bundling of sales.[840]

4.1.4 The "Cloud" Model

The possibility exists that the current "TPMs & contract" model will upgrade into a new "cloud" model. In this model, users do not need to download or store content on hard discs any more. As long as there is an Internet connection, they can access the content stored in the cloud at any time and at any place.

In recent years, the "cloud" model has emerged in cloud computing.[841] As Michael Miller points out:

> Cloud computing portends a major change in how we store information and run applications. Instead of running programs and data on an individual desktop computer, everything is hosted in the "cloud" – a nebulous assemblage of computers and servers accessed via the Internet. Cloud computing lets you access all your applications and documents from anywhere in the world, freeing you from the confines of the desktop and making it easier for group members in different locations to collaborate.[842]

The key factor necessary to understand the notion of cloud computing is the "cloud" itself. To put it simply, "the cloud is a large group of interconnected computers".[843] These computers belong to multiple companies or servers.

---

customer/display.html/ref=hp_mp3land_mp3land_mp3st_comp?nodeld=200455500&#compatible (last visited Mar. 4, 2011).

839 As will be analyzed in detail in the later section, after being sued for bundled sales, Apple Inc. began to sell TPMs-free music.

840 See *infra* section 4.2.3.

841 More details about cloud computing, see e.g., MICHAEL MILLER, CLOUD COMPUTING: WEB-BASED APPLICATIONS THAT CHANGE THE WAY YOU WORK AND COLLABORATE ONLINE (US: Que Publishing 2009); ANTHONY T. VELTE, TOBY J. VELTE & ROBERT ELSENPETER, CLOUD COMPUTING: A PRACTICAL APPROACH (US: McGraw-Hill Companies 2010); CLOUD COMPUTING: PRINCIPLES, SYSTEMS AND APPLICATIONS (Nick Antonopoulos & Lee Gillam eds., London: Springer 2010); HALEY BEARD, CLOUD COMPUTING BEST PRACTICES FOR MANAGING AND MEASURING PROCESSES FOR ON-DEMAND COMPUTING, APPLICATIONS AND DATA CENTERS IN THE CLOUD WITH SLAS (Emereo Publishing 2008); JOHN W. RITTINGHOUSE & JAMES F. RANSOME, CLOUD COMPUTING: IMPLEMENTATION, MANAGEMENT, AND SECURITY (Boca Raton: CRC Press 2010).

842 MILLER, *id.* at 7.

843 *Id.* at 9.

Authorized users can access the stored data via the Internet anywhere in the world. Meanwhile, the underground infrastructure holding the cloud is basically invisible to users.[844] A cloud computing solution is made up of three components – clients, distributed servers, and the datacenter.[845]

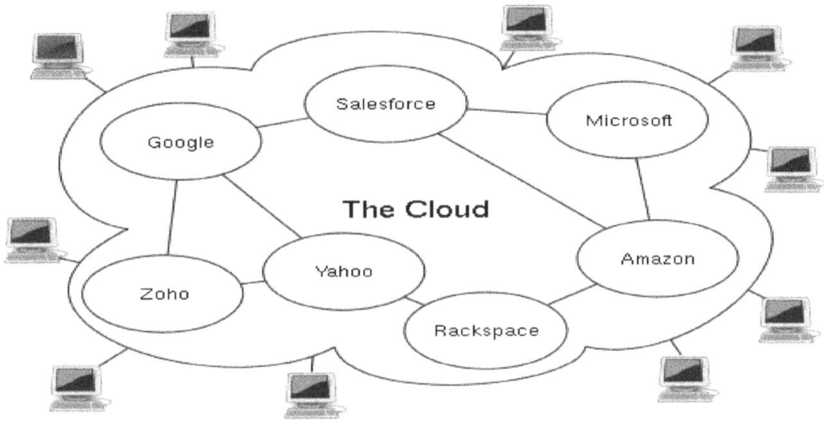

Fig. 4.2: Cloud Computing[846]

Conventionally, desktop computer users store their documents on personal computers. Despite the fact that documents can be accessed from other computers on a network, when your personal computer crashes, it is not available for others to use. By contrast, cloud computing users do not manage documents from their personal computers but rather access them through a collection of cloud servers via the Internet. Whatever content, documents, photographs, emails, or other data stored in the cloud become the users'. Therefore, unlike traditional PC-centric desktop computing, cloud computing is document-centric, task-centric, and user-centric.[847]

844 For example, the cloud services are based on HTTP, XML, HTML, JavaScript and other computing technologies which are invisible to users.

845 See VELTE, VELTE & ELSENPETER, *supra* note 841, at 6.

846 Quoting the picture from http://www.thetechscoop.net/2010/02/28/cloud-computing-defined/#axzz1KDR39tWY (last visited Mar. 4, 2011).

847 See MILLER, *supra* note 841, at 8-10.

At present, cloud computing is changing the way we use our computers and the Internet and has been attracting more and more attention.[848] One of the most popular cloud-based businesses is the Google cloud applications, including Gmail, Picasa, Google Docs & Spreadsheets, Google Calendar and so on. Not only Google, but also other companies are competing to occupy this new market:

> Big names that you know are jockeying for position in the cloud. Amazon, for example, offers several cloud sources from its Elastic Compute Cloud (EC2) to its Simple Storage Services (S3), which is part of its Amazon Web Services (AWS). Google has also put a number of applications online. At a basic level they are free, but more robust applications can be accessed with a subscription. Microsoft is betting big with massive new data centers to service millions of cloud users.[849]

As Philip Orbeta comments, "Cloud computing, though relatively new, is slowly starting to have a crucial impact on the way we handle information and Internet resources. It is possible that we may soon see cloud computing implemented as a standard in the near future.... [It] could become a gateway to a new age of information technology."[850]

4.2 No Best Alternative: Pros and Cons of Diverse Business Models towards Private Use

Pursuant to public choice theory, all institutions, to some degree, fall under the cloud of disrepute.[851] Of the various business models of digital content provision,

---

848 VELTE, VELTE & ELSENPETER, *supra* note 841, at 24 ("Compute clouds allow access to highly scalable, inexpensive, on-demand computing resources that run the code that they're given.").

849 *Id.* at xiv-xv.

850 Philip Orbeta, *How Things Work: Cloud Computing*, Mar. 1, 2010, available at http://thetartan.org/2010/3/1/scitech/cloudcomputing; more detail about the development of cloud computing, available at http://cloudcomputing.sys-con.com/ (last visited Mar. 10, 2011).

851 See David Skeel, *Public Choice and the Future of Public-Choice-Influenced Legal Scholarship*, 50 VAND. L. REV. 647, 659-60 (1997). More details about public choice theory, see generally e.g., DANIEL A. FARBER & PHILIP P. FRICKEY, LAW AND PUBLIC CHOICE: A CRITICAL INTRODUCTION (US: The University of Chicago Press 1991); Elhauge, *supra* note 53; Jerry Mashaw, *The Economics of Politics and the Understanding of Public Law*, 65 CHI.-KENT L. REV.123 (1989); MAXWELL STEARNS, PUBLIC CHOICE AND PUBLIC LAW: READINGS AND COMMENTARY (Cincinnati: Anderson Pub. Co., 1997).

each has its pros and cons to some extent, especially in respect of the issue of private use. Indeed, there may be no best alternative.

Table 4.2: Pros and Cons of Diverse Business Models

Model	Pros	Cons
TPMs & Contract	Low cost of anti-piracy; discriminatory pricing	Threatening computer safety; weakening enjoyment; preventing sharing and communication
Levy System	Paying levy in exchange for ability to use freely	No discriminatory pricing; weak linkage between the presumption that all user would buy legitimate products and the fact of piracy and P2P file sharing
Device-centric	Open source; ability to use freely	Bundling sale with device; preventing place-shifting and harmonization
Cloud	Convenient for exploitation & enjoyment	Access control problem; effect on subsequent creation

Table 4.3: How Private Use Is Being Facilitated or Prevented by Various Business Models

Model	Pros	Cons
TPMs & Contract	discriminatory pricing	Threatening computer safety; weakening enjoyment; preventing sharing and communication
Levy System	Paying levy in exchange for ability to use freely	No discriminatory pricing
Device-centric	Open source; ability to use freely	Bundling sale with device; preventing place-shifting and harmonization
Cloud	Convenient for enjoyment	Access control problem; effect on subsequent creation

4.2.1 The "TPMs & Contract" Model

The prominent advantages of the "TPMs & contract" model are the low cost of anti-piracy and discriminatory pricing. By virtue of TPMs' access control and price discrimination, users only need to pay certain fees depending on various uses of the works. This is more reasonable and affordable for both right holders and users. Since I have discussed the merit of TPMs control in previous sections, [852] here I will focus on its flaws in practice from the users' perspective.[853]

First, TPMs might threaten consumers' computer safety. TPMs "had several functions, which, in addition to limiting the number of copies a user could make, invaded the privacy of users by reporting information to the vendor and further created a serious security risk by installing undisclosed files on users' computers."[854] For example, by installing a "flawed and overreaching" computer program in sold music CDs, Sony BMG drifted into a scandal over threats it posed to consumers' computer security and privacy. As the Electronic Frontier Foundation (EFF) indicates:

> By including a flawed and overreaching computer program in millions of music CDs sold to the public, Sony BMG has created serious security, privacy and consumer protection problems that have damaged music lovers everywhere. At issue are two software technologies – SunnComm's MediaMax and First4Internet's Extended Copy Protection (also known as XCP) - which Sony BMG claims to have placed on the music CDs to restrict consumer use of the music on the CDs but which in truth do much more, including reporting customer listening of the CDs and installing undisclosed and in some cases hidden files on users' computers that can expose users to malicious attacks by third parties, all without appropriate notice and consent from purchasers. The CDs also condition use of the music on unconscionable licensing terms in the End User Licensing Agreement (EULA).[855]

After a number of class action lawsuits were filed against Sony BMG in many areas in the world,[856] Sony BMG compensated purchasers of CDs with the XCP

---

852 See *supra* sections 3.3.3 & 3.5.2.1.

853 Criticisms of TPMs control were discussed in sections 3.4 & 3.5.2.2 above, mainly from a theoretical perspective, with regard to free speech, dynamic efficiency, and users' fundamental rights angles.

854 Koren, *supra* note 48, at 1155, note 32.

855 EFF, *Sony BMG Litigation Info*, available at http://www.eff.org/cases/sony-bmg-litigation-info (last visited Mar. 10, 2011).

856 For instance, United States District Court for the Southern District of New York has dealt with many cases in the United States. See In re: Sony BMG CD Technologies Litigation, Case No. 1:05-cv-09575-NRB, available at http://www.eff.org/files/filenode/Sony-BMG/sony_prelim_approval_motion.pdf (last visited Mar. 11, 2011).

technology or the MediaMax technology, and ultimately stopped attaching these harmful TPMs on its CDs.[857]

Second, TPMs controls might weaken users' enjoyment. For instance, a popular file format – PDF – enables authors to exercise control over who is able to view, edit, copy or print the distributed PDF documents. These TPMs include password security, printing restrictions, copy text and graphic restrictions, modification or editing restrictions, adding or modifying comments, adding, removing, or rotating pages and so on.[858] Users, however, lose their freedom to manipulate the files.

Last but not least, TPMs controls prevent sharing and communication. A typical example is the region codes of DVDs. The DVD Copy Control Association (DVD CCA) requires that all DVD player manufacturers incorporate the Regional Playback Control (RPC) system.[859] The RPC system divides the world into six regions: ① North America; ② Western Europe, South Africa, Middle East and Japan; ③ Southeast Asia; ④ Australia, New Zealand and South America; ⑤ Africa and Russia; ⑥ China. Accordingly, DVDs distributed in each region can only be played on a player purchased in the same region.[860]

"Movie studios don't release each movie simultaneously worldwide. Instead, they release the same movie at different times in different regions, sometimes tailoring a movie's content and price for a specific part of the world."[861] Through DVD regional code playback controls, the copyright industries seek to maximize their economic interests while neglecting consumers' convenience in enjoying their products. As the Australian Competition & Consumer Commission (ACCC) commented:

> The practical effect is that a consumer who has bought a DVD player in Australia may be prevented from films obtained overseas. Overseas markets give Australian consumers access to a wide range of competitively priced film tiles, with special features not otherwise available in Australia.... this means Australian consumers are

857 For a further discussion of the Sony BMG incident, see Deirdre K. Mulligan & Aaron K. Perzanowski, *The Magnificence of the Disaster: Reconstructing the Sony BMG Rootkit Incident*, 22 BERKELEY TECH. L. J. 1157 (2007).

858 See Adobe PDF security measures, available at http://www.adobe.com/products/acrobatpro/protect-pdf-files-documents.html (last visited Mar. 10, 2011).

859 See JIM TAYLOR, DVD DEMYSTIFIED 187-190 (2d ed. NY: McGraw-Hill Companies, Inc. 2000);

860 See Bechtold, *supra* note 832, at 628-29.

861 ANDY RATHBONE, PCS: THE MISSING MANUAL 313 (CA: O'Reilly Media, Inc. 2006).

> forced to pay higher prices for films with fewer features and a more limited range of titles.[862]

Similar concerns have been expressed by the European Commission during investigations into whether the regional code playback controls in DVD players were used to overcharge European customers for DVD discs compared to US customers.[863] Furthermore, there is the additional concern that "regional code management systems can undermine the free movement of goods which intellectual property law protects by the exhaustion principle."[864]

4.2.2 The "Levy System" Model

In contrast with the "TPMs & contract" model, a distinct advantage of the "levy system" model is that it gives users the ability to use the works freely in exchange for the payment of a levy. The non-commercial use and sharing of cultural expression is a fundamental form of speech and not theft, and should thus be celebrated rather than suppressed. In this sense, the levy system serves as a middle way between "digital abandon", in which noncommercial uses of digital materials are free from copyright controls, and "digital lock-up", in which copyright industries control every use of and access to digital works.[865] Without the annoying TPMs restrictions, users can manipulate the copyrighted works in their favored manner. They can either merely consume the works or creatively participate by making comments, editing, or abridging, or making private copies and sharing them with family members and close friends.

However, non-discriminatory pricing has raised criticisms. As Hugenholtz indicates, imposing copyright levies on the sale price of multi-purpose devices,

---

862 Australian Competition & Consumer Commission, *ACCC Consumer Express*, p.1, Feb. 2002, available at http://www.accc.gov.au/content/index.phtml/itemId/302673 (last visited Oct. 8, 2010).

863 See *Answer of the European Commission to Written Questions E-1509/00 and E-1510/00*, Official Journal of the European Communities C 53 E, Feb. 20, 2001; *Answer of the European Commission to Written Questions E-2371/00*, Official Journal of the European Communities C 103 E, Apr. 3, 2001; Australian Competition & Consumer Commission, *ACCC Consumer Express*, Feb. 2002, available at http://www.accc.gov.au/ content/index.phtml/itemId/302673 (last visited Oct. 8, 2010).

864 Bechtold, *supra* note 832, at 629 (supplying that although the regional code management system seemed to be larger than the domestic geographical areas in which IPRs become exhausted, this might be a problem if the principle of international exhaustion would apply).

865 See Netanel, *supra* note 6, at 83.

such as personal computers, radio and TV sets, digital cameras, and telephones may result in an unjustified and undesirable levying on the copyright system.[866]

Traditionally, copyright levies are targeted against devices whose primary use is to reproduce works, for instance, tape recorders and blank tapes for recording music, or photocopying machines for copying written works. However, in the digital environment, imposing levies on a wide range of digital equipment, devices or media whose primary purpose is not for copying, would indiscriminately affect all users regardless of whether they copy or not. Consequently, levy-targeted digital goods and services are charged higher fees which distort market and consumer demand.[867]

4.2.3 The "Device-Centric" Model

In the "device-centric" model, the attraction of digital content is their cheap price, open source and lack of TPMs. However, there is no such thing as a free lunch. The whole distribution, storage and usage of digital copyrighted materials heavily rely on or are constrained by the device. For example, as I mentioned above, the iTunes music exists to drive sales of the companies' expensive music players. In order to do so, bundling sales with devices, preventing place-shifting and harmonization are inevitably included with this business model.

Although there are over 13 million TPMs-free songs priced at just 69 ¢, 99¢, or $1.29 each, songs purchased from iTunes are formatted with FairPlay, which prevents consumers from playing songs on other devices.[868] Similarly, songs sold on Sony's online music store – CONNECT – are downloadable in ATRAC3 format, which is a proprietary format that can only be played on Sony players.[869]

In addition to preventing the place-shifting of iTune music, Apple Inc. also refuses to provide licenses to companies seeking compatibility, and even

866 See MAZZIOTTI, *supra* note 510, at 257.

867 See *id.*

868 See John Borland, *Apple Fights RealNetworks' "Hacker Tactics"*, CNET News, Dec. 14, 2004, available at http://news.cnet.com/Apple-fights-RealNetworks-hacker-tactics/2100-1027_3-5490604.html ("Songs purchased from Apple's iTunes store can only be played directly on an Apple iPod".) (last visited Apr. 15, 2011).

869 ATRAC stands for Adaptive Transform Acoustic Coding. The earliest ATRAC system – ATRAC*1 has evolved into ATRAC3, ATRAC3plus and ATRAC Advanced Lossless. ATRAC Audio Device allows users to download and play music purchased from Mora in Japan, and CONNECT in North America and Europe. For more details about ATRAC, available at http://www.sony.net/Products/ATRAC3/index.html (last visited Apr. 15, 2011).

prevents other music distributors from providing music harmony with Apple players. In one notable example, in 2004, RealNetworks released harmony software which enabled songs sold from the company's online store be played on the iPod.

Several record labels executives praised RealNetworks' effort to achieve compatibility with the iPod without Apple Inc.'s permission. Traditionally, CDs and DVDs have worked on any manufacturers' players while music online providers have been tied to specific brands of devices. Thus, "[w]e applaud RealNetworks' efforts to help correct this situation and appeal to all people and companies in this area to work toward a world of universal interoperability," said Larry Kenswil, president of Universal Music's eLabs division.[870]

However, Apple Inc. criticized RealNetworks' harmony software as being the tactics of a hacker, and threatened legal action under the DMCA which broadly restricts the circumvention of TPMs, to force RealNetworks give up its compatibility formats.[871] In addition, Apple Inc. quietly updated its iPod software so that songs purchased from RealNetworks' online store could no longer play on it.[872]

In response, "Label executives ... howled about Apple's digital rights management, FairPlay, which limited iTunes-purchased tracks to the iPod rather than worthy competitors like Creative Labs's Zen or Microsoft's Zune.... [P]ressure [also came] from European regulators, who in the late 2000s started to accuse Apple of illegally cornering the digital music market."[873]

It is not only record labels and competitive rivals who are against Apple Inc.'s bundling sales, consumers are also dissatisfied with the artificial restriction on the use of works. For instance, an American customer raised a class action against Apple Inc. on July 21, 2006 in the United States District Court for the Northern District of California. The plaintiff argued that Apple failed to give consumers notice that music purchased from the iTunes Store was incompatible with music players provided by other vendors. The plaintiff further argued that Apple Inc.'s behavior of unlawfully tying music purchased on the

870 See John Borland, *RealNetworks Breaks Apple's Hold on iPod*, CNET News, July 26, 2004, available at http://news.cnet.com/RealNetworks-breaks-Apples-hold-on-iPod/2100-1027_3-5282063.html?tag=mncol (last visited Apr. 15, 2011).

871 See Matt Hines, *"Stunned" Apple Rails Against Real's iPod Move*, CNET News, July 29, 2004, available at http://news.cnet.com/Stunned-Apple-rails-against-Reals-iPod-move/2100-1041_3-5288378.html?tag=mncol (last visited Apr. 16, 2011).

872 See Borland, *supra* note 868.

873 KNOPPER, *supra* note 836, at 180.

iTunes Store with the purchase of iPods in order to maintain a monopoly market power violated antitrust laws.[874]

4.2.4 The "Cloud" Model

Compared to other business models of digital content provision, "cloud computing is considered by many to be the next paradigm shift in computer technology, and this may be the next 'platform' for innovative ecosystems."[875] "The basic idea behind cloud computing is that anything that could be done in computing – whether on an individual PC or in a corporate data center – from storing data to communicating via email to collaborating on documents or crunching numbers on large data sets – can be shifted to the cloud."[876] As the Open Computing Manifesto points out, "[t]he key characteristics of the cloud are the ability to scale and provision computing power dynamically in a cost efficient way and the ability of the consumer (end user, organization or IT staff) to make the most of that power without having to manage the underlying complexity of the technology".[877]

When digital contents are stored in the cloud, users no longer need to download or store them on hard disc or portable devices. As long as there is an Internet connection, they can access the content stored in the cloud at any time and at any place. This model would greatly facilitate users' enjoyment of digital works. Take the example of a consumer who previously purchased a Microsoft Word CD-or DVD-ROM and installed it on their personal computer. Every time Microsoft issued a new version thereafter, instead of buying a new CD the consumer can now simply go to a website, pay a small fee (if any) and click the online upgrade button.

On the other hand, when there is no necessity or demand for copying, the reproduction right which is the traditional core of the copyright system might be inadaptable to digital cloud computing. As Zohar Efroni indicates:

---

874 See *Tucker v. Apple Computer, Inc.*, 2006 U.S. Dist. LEXIS 96343 (N.D.Cal. Dec. 20, 2006).

875 VELTE, VELTE & ELSENPETER, *supra* note 841, at 67.

876 David C. Wyld, *Risk in the Clouds?: Security Issues Facing Government Use of Cloud Computing*, in INNOVATIONS IN COMPUTING SCIENCES AND SOFTWARE ENGINEERING 6 (Tarek Sobh & Khaled Elleithy eds., Heidelberg: Springer 2010).

877 *Open Cloud Manifesto: Dedicated to the Belief That the Cloud Should be Open*, OpenCloudManifesto.org, Spring, 2009, available at http://opencloudmanifesto.org/opencloudmanifesto2.htm (last visited Apr. 17, 2011).

> The reproduction right belongs to past chapters in copyright law's evolution; it has grown evidently unsuitable to usher the system into the digital future. The notion of fixation is structurally inimical to contemporary modes and methods of providing and gaining access to digital information. Fixation, in other words, is the 'yesterday' of copyright law.[878]

Consequently, some commentators argue that exclusive copyright tailored around the concept of fixation or reproduction cannot continue to be the body of digital copyright law.[879] Instead, the digital communication right as an access-right should be a crucial component in the digital copyright regime.[880] When everything is stored in a cloud, access to the cloud becomes a precondition of using the stored content. In order to secure the safety of digital contents in the cloud, TPMs on access control[881] are a prerequisite.[882]

However, the digital lock-up of the copyrighted information might generate access control problems, and suppress subsequent creativity from a dynamic efficiency perspective. Given the fact that copyrighted content is both an input and an output of the creativity process, the perfect control on access as a strategy for generating revenues would raise the cost of secondary creativity, and decrease the incentives of the subsequent creators.[883] Moreover, as Oliver Goldsmith indicates, "Laws grind the poor, and rich men rule the law."[884] If the rich in an economy have access, and the poor are excluded, the cultural Matthew effect generated by access control might harm a sound democratic culture.

---

878 ZOHAR EFRONI, ACCESS-RIGHT: THE FUTURE OF DIGITAL COPYRIGHT LAW 247-48 (NY: Oxford University Press 2011).

879 See Paul Edward Geller, *Beyond the Copyright Crisis: Principles for Change*, 55 J. COPYRIGHT SOC'Y U.S.A. 165, 179-81 (2008); Ernest Miller & Joan Feigenbaum, *Taking the Copy Out of Copyright*, available at http://cs-www.cs.yale.edu/homes/jf/MF.pdf (last visited Apr. 18, 2011).

880 See EFRONI, *supra* note 878, at 284-85.

881 See CHARLES P. PFLEEGER & SHARI LAWRENCE PFLEEGER, SECURITY IN COMPUTING 7 (3rd ed., New Jersey: Pearson Education, Inc 2003) ("We use a control as a protective measure. That is, a control is an action, device, procedure, or technique that removes or reduces a vulnerability.").

882 See *id.* at 326 ("The challenge of the access control problem is to limit users' access so that they can obtain only the data to which they have legitimate access.").

883 See *supra* section 3.5.2.

884 OLIVER GOLDSMITH, THE TRAVELER, line 386 (1764), quoting GEORGE MILLER, ON FAIRNESS AND EFFICIENCY: THE PRIVATISATION OF THE PUBLIC INCOME DURING THE PAST MILLENNIUM 330 (UK: The Policy Press 2000).

4.3 Business Models and Required Social Conditions

According to the institutional analysis approach to law, each institution has its merits and flaws to some extent. It is impossible to find the "best" institution, rather the search is for the least imperfect alternative.[885] Consequently, the focus of this paper is not to determine which of the aforementioned business models is the best, but rather to determine which social conditions need to be met in order for each of them to be effective.

Table 4.4: Social Conditions of Various Business Model

Business Model	Social Conditions Should be Met in Order to Be Effective
TPMs & Contract	Deleting TPMs in certain cases; third-party institution to deal with the deletion and supervise the proper use of TPMs deleted works
Levy System	Limit to equipment, device, or media whose primary purpose is to reproduce works; institution to collect and manage levy
Device-centric	Preventing bundled sale with devices; allowing place-shifting and harmonization
Cloud	Full penetration of Internet; infrastructure to promote access to knowledge and creation-enhancing

4.3.1 The "TPMs & Contract" Model

First, in order to restore the delicate balance between stakeholders and overcome the drawbacks of perfect control by TPMs, it is necessary to allow for the deletion of TPMs in certain cases for the sake of fair use and other copyright limitations. As Ian Kerr points out:

> [I]f the copyright industries and national governments insist that there is a legitimate need for new laws to prevent the circumvention of TPMs, then similar provisions are needed to protect citizens from organizations that use both TPMs and the law of contract as a kind of privacy circumvention device. Copyright owners should not be

885 See KOMESAR, IMPERFECT ALTERNATIVES, *supra* note 51, at 3-13.

encouraged or allowed to use TPMs and contracts to circumvent fair information principles.[886]

In addition, it is not always copyright infringement in the case of the unauthorized circumvention of TPMs. We need to separate legitimate TPMs for copyright protection from outstretched TPMs which are merely designed to maintain a market monopoly. For instance, Sony Inc. sued the manufacturers of "mod chips" which circumvented region codes, and allowed users of Sony PlayStations to play games purchased in different regions. The Australia High Court distinguished between pirating a game and playing with legitimate copies using a mod chip. The High Court held that although making a pirated copy of a game is illegal, playing legitimate copies using a mod chip is not. Furthermore, the circumvented regional code of Sony Inc. did not protect a copyright, but intentionally reduced global market competition and limited consumers' rights.[887]

Finally, third-party institutions are needed to deal with the deletion and to supervise the proper use of works after TPMs have been deleted. A third party institution is preferable to the use of a collective management organization or consumer organization for this purpose as it will be less prone to accusations of bias and will thus be better able to obtain the trust of the public.[888]

Generally speaking, there are two approaches for institutional design – building new institutional bodies or using the existing ones. In practice, either administrative agencies or judicial bodies are chosen for this new function. First, a special governmental or judicial institution is set up to deal with the conflicts between copyright limitations and technological measures. For example, Finland established a special arbitration body to tackle disputes between right holders and users.[889] In 2007, France created the Regulatory Authority for Technical Measures (l'Autorité de Régulation des Mesures Techniques or ARMT), an

886 Ian Kerr, *To Observe and Protect? How Digital Rights Management Systems Threaten Privacy and What Policymakers Should Do about It*, in INTELLECTUAL PROPERTY AND INFORMATION WEALTH: ISSUES AND PRACTICES IN THE DIGITAL AGE, VOL. 1, COPYRIGHT AND RELATED RIGHTS 322 (Peter K. Yu ed., Connecticut: Praeger Publishers 2007).

887 See *Stevens v. Kabushiki Kaisha Sony Computer Entertainment* [2005] H.C.A. 58, paras. 101-104 (Austl.).

888 See e.g., Engelbrert, *supra* note 173, at 75 ("collective management organizations …with unmistakable allegiance to the cause of right-holders").

889 See Guido Westkamp, *The Implementation of Directive 2001/29/EC in the Member States*, Part II, IVIr, Report Commissioned by the European Commission DG Internal Market, Feb. 2007, p.68, available at http://www.ivir.nl/publications/guibault/InfoSoc_Study_2007.pdf (last visited Apr. 18, 2011).

independent regulatory agency charged with promoting the interoperability of digital information distributed with embedded TPMs.[890] As Antonina Engelbrert explains:

> The Authority is empowered to rule on any conflict between exceptions and technological measures. It has the general competence to ensure that the exceptions will be observed and to determine the way the exceptions should be respected in applying the TPMs, as well as the number of copies that should be made possible. The ARTM acts upon request of any beneficiary of a relevant exception or of an association...and generally should strive to facilitate conciliation between parties.... The ARTM has the status of an independent administrative authority.... It is composed by way of a government decree and consists of six members.... The members are appointed for a time-limited mandate of six years which is non-revocable and non-renewable. Apparently, this composition seeks to provide authority, expertise and integrity, rather than representation of affected interests. In this, the Authority is much more a regulatory than a self-regulatory or a co-regulatory body. When a case has been referred to the Authority, it seeks to achieve conciliation. Upon failure to reach agreement within two months the ARTM must either decline the request or issue a prescriptive injunction, possibly upon penalty of a fine... The decision of the Authority may be appealed before the Paris Court of Appeal with a suspensive effect.[891]

The second approach is to make use of existing institutional bodies. For instance, the existing Copyright License Tribunals in Norway and Denmark have been endowed with new functions to deal with TPMs related claims. Take the Danish Copyright License Tribunal as an example, upon request, the Tribunal may order right holders who employ effective TPMs on their works to provide available means for users in order to exert copyright limitations. In case the right holders do not follow the order within 4 weeks of the Tribunal making a decision, users may circumvent the TPMs without liability for violating anticircumvention rules.[892]

4.3.2 The "Levy System" Model

In my opinion, there are at least two requirements for copyright levy scheme to be effective. First, the charged copyright levies should be reasonably priced.

890 See Nicolas Jondet, *DRM Watchdog Established in France* (Décret n 2007-510 du 4 Avril 2007), French-law.net, Apr. 11, 2007, available at http://french-law.net/drm-watchdog-established-in-france-decret-2007-510-4-avril-2007.html (last visited Apr. 12, 2011).

891 Antonina B. Engelbrert, *supra* note 173, at 88-89.

892 See Westkamp, *supra* note 889, at 166.

Second, copyright societies or collecting societies should be responsible for collecting, managing and distributing the levies. In this section, my analysis will focus on the first condition which is facing great challenge in the digital environment.

Basically, the pricing of levies is evaluated by the supply and demand in the market. However, a fundamental principle is that copyright levies are and should be limited to equipment, devices, or media whose primary purpose is to reproduce works.[893]

In a report entitled *Copyright Issues in Digital Media*, the United States Congressional Budget Office pointes out that charging copyright levies on computers or other general-purpose digital equipment was not desirable:

> From an equity point of view, levying a fee on computers and other equipment used to access certain creative works could also impose a tax on the use of that equipment for activities unrelated to the copyrighted materials in question. For example, taxing purchases of computers or other digital media to compensate recording artists for Internet downloads would force individuals who use computers for other purposes to subsidize the online music consumption of others.[894]

Similarly, Verica Trstenjak, Advocate General at the EU Court of Justice, believes the levy imposed on digital equipments should be limited to cases "where it may be presumed that they are to be used for private copying".[895] Spain allows the unauthorized reproduction of published works for private use in exchange for lump-sum compensation by means of a levy paid by manufacturers, importers or retailers to IPRs management societies on private copies. Applied indiscriminately to digital reproduction equipment, devices and media, Trstenjak suggests that such a levy may not be applied indiscriminately to those who acquire equipment and data media for other purposes.[896]

The underpinning justification for the imposition of levies on digital equipment, devices or media is the contributory liability rationale. A statutory levy is legitimate from a users' perspective only when adequate causation exists "between the act of making a reproduction of a copyrighted work, the media or equipment used to make the reproduction and the harm done to the interests of the right owner".[897] The GEMA cases, decisions of the German Supreme Court

---

893 See Guibault, *supra* note 825, at 242.

894 CONGRESSIONAL BUDGET OFFICE, COPYRIGHT ISSUES IN DIGITAL MEDIA (Washington, D.C.: CBO 2004).

895 ECJ Advocate General: Limitations to the levy on private copies, May 19, 2010, available at http://www.edri.org/edrigram/number8.10/ecj-levy-digital-equipment (last visited Apr. 20, 2011).

896 See *id.*

897 Guibault, *supra* note 825, at 242.

(BGH), emphasized the point that levy schemes should be based on the contributory liability of equipment manufacturers and traders.[898]

Therefore, levies on multipurpose equipment, such as PCs, hard disks, TV sets, digital cameras, and mobile phones, no longer reflect the contributory liability rationale on which the levy system is rooted. For instance, a PC is used for a wide range of activities, many of which are irrelevant from a copyright perspective, such as word processing and table drawing. Levies on general purpose devices would "penalize those who do not infringe copyright and may encourage copyright infringement amongst consumers who consider that they have already paid for the right to infringe copyright".[899] Lawmakers should be more cautious about an expansive copyright levy imposed on general digital equipment or media. As a report entitled *The Future of Levies in a Digital Environment* points out:

> In the end, an increasingly large number of users would end up paying a "copyright tax" without actually using copyrighted content. Large numbers of consumers would be cross-subsidising a relatively limited group of "private copiers." Moreover, such an all-encompassing levy scheme would be perceived by many users as an "unlimited license to copy". We predict that such an expansion would eventually undermine the copyright system as a whole. Following a near-total "levitation" of the copyright system, exclusive rights would effectively cease to exist. Right holders would instead become totally dependent on remuneration rights collected by collecting societies.[900]

898 See *id.*; The German Society For Musical Performing Rights And Mechanical Reproduction Rights (GEMA) is the German collecting society which represents the copyrights of more than 64,000 members, and over 2 million copyright. For more details about GEMA, see https://www.gema.de/en/home.html (last visited Apr. 20, 2011).

899 Nicolas Blackmore, *Peer-To-Peer Filesharing Networks: The Legal and Technological Challenges for Copyright Owners*, New South Wales Society for Computers and the Law, available at https://www.nswscl.org.au/index.php?option=com_content&view=article&id=124:peer-to-peer-filesharing-networks-the-legal-and-technological-challenges-for-copyright-owners&catid=27:march-2004-issue&Itemid=31 ("They may either discourage copyright owners from pursuing chronic copyright infringers with legal action or provide copyright owners with a windfall when they succeed against such infringers".) (last visited Apr. 15, 2011).

900 Bernt Hugenholtz, Lucie Guibault & Sjoerd van Geffen, *The Future of Levies in a Digital Environment*, Final Report, Institute for Information Law, Amsterdam Mar. 2003, p.3, available at http://www.feei.at/img/db/docs/327.pdf (last visited Apr. 16, 2011).

4.3.3 The "Device-Centric" Model

According to the aforementioned analysis, the bundling of sales with devices, the prevention of place-shifting and harmonization are the main shortcomings of the "device-centric" model. Therefore, preventing bundled sale with devices, and allowing place-shifting and harmonization are the necessary social conditions to be met. In practice, some content provision and device reproduction companies have already begun to change their strategies out of market or litigation concerns.[901]

For instance, after being sued for anti-competitive behavior in relation to bundled sales, Apple Inc. announced in 2009 at the Macworld Conference & EXPO that by April 2009, all songs from iTune would be TPMs free. iTune music can now be transfered into AAC files used on other players.[902] In addition, since 2009, iTune9 has been compatible with Windows 7, and the latest version iTunes 10 is compatible with Windows XP and later versions.

Harmonization is not merely a case of Apple bowing to the pressure from Windows or others for the ability to install iTunes software, it is also a profitable marketing strategy for Apple. For example, by virtue of its compatibility with Windows, Apple Inc. attracts consumers who use Windows rather than the Mac to use their iPod. This is extremely important since most computer users are using the Windows operation system rather than Mac. In this way, Apple Inc. is exploring and expanding to new markets.

Notably, Apple Inc. has always adopted an independent format distinct from other providers. Although it is now technologically possible to employ software to transfer the format and play the content on other players, this process is notoriously inconvenient. Some additional functions cannot be exerted after the transfer as well. Thus, in order to optimize their experience, consumers prefer to

---

901 E.g., Sony's online CONNECT Music Store sold ATRAC-encoded files only playable on Sony devices. Due to its lack of success in catching on with consumers in the North American and European markets, on Aug. 30, 2007, Sony announced plans to phase out the ATRAC-only CONNECT Music Store after Mar. 2008 in the west, while still maintains its Japanese equivalent Mora. The two new Walkman players, the NWZ-A810 and NWZ-S610 use Microsoft's Windows Media Audio format, and play MP3 and AAC files, while not Sony's proprietary ATRAC format. See Sony CONNECT Europe, *Important Information*, available at http://www.connect-europe.com/; Michael Calore, *Sony Ditches ATRAC, Connect Music Store*, available at http://www.wired.com/gadgets/portablemusic/news/2007/08/atrac; Rob Pegoraro, *Sony Connect, ATRAC Exit the Stage*, available at http://voices.washingtonpost.com/fasterforward/2007/08/sony_connect_atrac_exit_the_st.html (last visited May. 15, 2011).

902 The TPMs free approach is, however, only limited to music. Videos and TV shows sold on iTune still retain TPMs.

purchase a full set of Apple products, such as the Mac laptop, iPod, and iPhone. In this sense, the bundling sale of contents with devices has stealthily shifted from "hard" to "soft".

4.3.4 The "Cloud" Model

First of all, on account of the fact that cloud computing is based on providing services on demand from a remote location over the Internet,[903] the prevalent application of the "cloud" model to a great extent depends on the sufficient penetration of the Internet. In other words, the physical precondition of cloud computing is Internet access. Without a broadband Internet connection, users can not store their data in cloud servers or access their data at any time and place. Therefore, the penetration of the Internet will determine the development and application of cloud computing in a given country or area.

In respect of the aforesaid limitations of access controls in the "cloud" model on subsequent creativity, certain infrastructure is needed in order to promote access to knowledge and creativity. In fact, the access control problem has also existed in the "TPMs & contract" model. As noted above, a third-party institution is required to deal with the deletion of TPMs in certain cases, and supervise the proper use of TPMs deleted works.[904] Thus, a possible way of constructing the infrastructure needed to balance access and control in the "cloud" model is to introduce a third-party institution that would allow access in certain cases, and supervise the subsequent proper use after access to the cloud has been granted.

4.4 Problems in Chinese Practice of Reaching the Necessary Social Conditions

"Institutional choice is contingent on a historical and institutional context that has been shaped through time, is often country-specific and is generally resistant to change."[905] When reaching the same social conditions, different flaws might

---

903 See Wyld, *supra* note 876, at 6 ("The term 'cloud computing' has at its core a common element – in that with the cloud model, computing services are delivered over the Internet, on demand, from a remote location, rather than residing on one's desktop, laptop, mobile device, or even your own organization's servers.").

904 See *supra* section 4.3.1.

905 Engelbrekt, *supra* note 173, at 71.

occur in various countries on account of the realistic situations. In this section, I intend to explore the problems in Chinese practice.

Table 4.5: Social Conditions and Their Flaws in Chinese Practice

Model	Social Conditions	Flaws in Chinese Practice
TPMs & Contract	Deleting TPMs in certain cases; third-party institution to deal with the deletion and supervise the proper use of TPMs deleted works	Difficult to distinguish between private use and piracy; agency capture & corruption
Levy System	Limit to equipment, device, or media whose primary purpose is to reproduce works; collective society to collect and manage levy	
Device-centric	Preventing bundling sale with devices; allowing place-shifting and harmonization	Conflict with commercial profits
Cloud	Full penetration of Internet; infrastructure to promote access to knowledge and creation-enhancing	Time consuming; security concerns; difficult to operate

4.4.1 The Difficulty in Distinguishing Between Private Use and Piracy

In order to maintain a balance between copyright holders' economic benefits and users' freedom of enjoyment, it is necessary to allow the deletion of TPMs in certain circumstances for the sake of copyright limitations in the "TPMs & contract" model. Similarly, in the "levy system" model, users are charged copyright levies in exchange for the ability to use the works freely. However, the precondition for doing that is separating private use from piracy. That is to say, in both the "TPMs & contract" model and the "levy system" model, lawful private use should be protected, while illegal piracy should be punished. The difficulty is how to separate them in practice.

As mentioned above, private use is full of uncertainty both in theory and in application. Given the fact that the copying itself is physically the same, it is almost impossible to set an absolutely clear boundary between private use and piracy, and a context-originated determination must always be made.[906] The difficulty in distinguishing between private use and piracy is extraordinarily apparent in China. On the one hand, there is no solid definition of private use in the Chinese copyright statue. The only prescription explicitly referring to private use is the private use exemption within fair use. Article 22 of the 2010 Chinese Copyright Law provides that any use as long as it is for the user's own private study, research, or self-entertainment purposes is exempted from copyright infringement. However, this broad conception of private use does not set up a clear boundary between private use and piracy, and thus has been abolished in the fair use scheme of digital copyright.[907]

On the other hand, the blurred distinction between enjoyment and exploitation in the digital environment has induced the sacrifice of the private use limitation. The abandonment of the private use limitation of digital copyright actually serves as an anti-piracy measure.[908] The confusion between private use and piracy has therefore existed in China for a long time. It won't be easy to remove all the uncertainty in both theory and in application.

4.4.2 Agency Capture & Corruption

In the cases of the "TPMs & contract" model and the "levy system" model, certain institutions are needed for the management. However, pursuant to the pessimism of public choice theory, all institutions – administrative agencies, legislatures and even the "insulated" courts – have to some degree, fallen under the cloud of disrepute.[909] Under this "capture pathology", both third-party institutions and copyright collective societies are thus subject to agency capture or corruption.

Due to the fact that collective societies and consumer organizations are suspected of harboring biases, the search for a third-party institution to deal with TPMs related copyright disputes must begin with independent third parties who could obtain the trust of the public. This third-party institution, however, might itself also be captured by interest groups. In light of Neil Komesar's theory on minoritarian bias, the concentrated high-stake interests of copyright holders

906 See *supra* section 3.1.

907 See *supra* sections 1.2 & 1.3.

908 See *supra* section 2.1.

909 See Skelly Wright, *The Courts and the Rulemaking Process: The Limits of Judicial Review*, 59 CORNELL L. REV. 375, 397 (1974).

often prevail over the dispersed small-stake interests of users in the legislative process as well as in public agency decision-making.[910]

Moreover, the situation might be more complicated in a Chinese context on account of the administrative dominant "dual system" and the multi-headed copyright administrative agencies involved in copyright protection affairs. The rival powers of these multi-headed administrative agencies might either cause stagnation and generate policy inconsistencies, or foment a minoritarian bias when one agency captures the legislature.[911] Therefore, if we set up a new agency or utilize an existing one to tackle TPMs related copyright disputes, a difficult problem is how to deal with its relationship with other institutions within the "dual-system".

In respect of collective societies, the distinct problems are lack of transparency and corruption. Among various collective societies, the Music Copyright Society of China and the China Audio-Video Copyright Association are the most important ones.[912] However, multiple problems, especially the lack of transparency and corruption in the collection, management, and distribution of copyright levies, undermine collective societies' reputation. Given the fact that right holder's interests cannot receive effective and sufficient protection, some copyright holders hesitate or have a lack of enthusiasm to become members of collective societies.

4.4.3 The Conflict Between Free Competition and Commercial Profit

The policies of the market economy include the principles of free competition in production, free consumer choice, and the free development of the personality.[913] Freedom of competition is the basis of any market economy. "This can only exist where nothing is permitted to suppress freedom, but where

910 See KOMESAR, IMPERFECT ALTERNATIVES, *supra* note 51, at 76. For more discussion about the interest contest between right holders and users, as well as the minoritarian bias phenomenon, see *supra* section 3.5.2.2.2.1.

911 See *supra* sections 2.2.2.2.1 & 2.2.2.2.2.

912 For more details about the Music Copyright Society of China, see http://www.mcsc.com.cn/ (last visited Apr. 15, 2011); information about the China Audio-Video Copyright Association, see http://www.cavca.org/ (last visited Apr. 15, 2011).

913 See LUDWIG ERHARD, PROSPERITY THROUGH COMPETITION 51 (NY: Frederick A. Praeger, Inc. 1958).

freedom is guaranteed by statute and is one of the most prized things in the community."[914]

Accordingly, the bundling of sales and lack of compatibility are opposed to the free competition requirements. Strictly following the principle of free competition, however, sometimes conflicts with companies' short-term commercial profits. As a profit maximizer, a firm might violate free competition, especially in the current environment in China with its unsound market economy.

A typical example was the fight between Tencent QQ and Qihoo 360 in 2010. QQ is a free chat and message software similar to MSN, and Qihoo 360 is a free software which was originally an antivirus program. The war started when Qihoo 360 alleged that QQ was leaking users' private data and offered a new service on 360 Safe to prevent privacy leaks. Tencent accused Qihoo 360 of slander and business foul play. "A new tag-on service run by 360 Safe has affected the normal functioning of QQ. With 360 Safe installed, we can not guarantee the safety of our service. To avoid our clients' computer desktops being turned into a battlefield, we have made this 'hard' decision." Tencent said in a statement popping up on the screens of millions of QQ and 360 Safe users. [915]

In fact, both of the two free software companies have developed advanced accounts payable services and become competitors in some services. In this battle, the two companies fought with each other by adding TPMs on their software to prevent the other's from being compatible on user's computer. If you installed Qihoo 360, you could not use QQ, and vice versa.

Both of them argued that given the fact that users are using their service for free, as copyright holders, they were legally allowed to control the use of their software through TPMs. However, the vicious competition was heavily criticized by consumers. According to an online survey conducted by Sina.com, over 78.1 % of participating Internet users said Tencent and Qihoo 360 fought for their own interests but both forgot the interests of their clients. As Liu Xingliang, an Internet commentator indicated, "[W]hen the populace is angry, neither company wins the war."[916]

Moreover, when there are hundreds of millions of people using QQ and Qihoo 360, the conflict was no longer a private issue merely between the two companies or between right holders and users, it upgraded into a public interest

914 *Id.* at 103.

915 See *Popular Messaging Service Shut Down as Fight Between China's Internet Firms Escalates*, Xinhua News, Nov. 4, 2010, available at http://news.xinhuanet.com/english2010/sci/2010-11/04/c_13589831.htm (last visited May 15, 2011).

916 See *id.*

incident.[917] In the end, after the intervention of the Ministry of Industry and Information Technology, the two companies finally returned their services to full compatibility.

This incident reminds us that the transition from a planned economy to a socialist market economy in China is a long-term and arduous task. Further efforts need to be made in order to reach the blueprint listed in Some Issues Concerning the Establishment of a Socialist Market Economics System.[918]

4.4.4 Full Internet Penetration Takes Time

The technological infrastructure of cloud computing is entirely depending on Internet access. In order to make sure that users can access the stored data in the cloud at any time and at any place, full penetration of the Internet is required. According to the worldwide Internet penetration statistics, however, there is a long way to go before that goal is reached.

In general, there is an imbalance in world Internet usage. The Internet penetration rates are much higher in developed countries than in developing countries. For instance, Japan's (79.5%) is more than 5 times higher than

917 According to the companies' websites, the number of QQ users in China has reached 1 billion, and the number of Qihoo 360 users is at least 300 million by June 2010. See *id.*

918 The document was approved by the Third Plenary Session of the Fourteenth Party Central Committee held in Nov. 1993. It drew the outline for establishing a new socialist market economy: "(1) Establish a socialist market economy means making the market a fundamental factor in the allocation of resources. To achieve this goal, it is necessary to keep public ownership as the mainstay, further transform the management mechanism of state-owned enterprises and establish a modern enterprise system that suits the requirements of a market economy. Such an enterprise system would clearly outline property rights, enterprise rights and responsibilities. (2) Establish a unified and open market network, closely combine urban and rural markets, link domestic and international markets and promote an optimal distribution of resources. (3) Transform government functions in managing the economy and establish a complete macro-control system mainly through indirect means. (4) Establish an income distribution system with the main principle of distribution according to work while giving priority to efficiency and stressing fairness, encourage some people in some area to get rich first. (5) Establish a multi-level social security system for urban and rural resident to promote the economic stability. This constitutes the basic framework of socialist market economy system that is interconnected." See Zhonggong zhongyang guanyu jianli shehui zhuyi shichang jingji tizhi ruogan wenti de jueding [DECISION OF THE CENTRAL COMMITTEE OF THE COMMUNIST PARTY OF CHINA ON SOME ISSUES CONCERNING THE ESTABLISHMENT OF A SOCIALIST MARKET ECONOMICS SYSTEM] 4 (Beijing: Renmin chubanshe [People's Publishing House] 1993).

Africa's as a whole (15.6%). Moreover, even within the same area, there is obvious discrepancy among different countries, such as the gaps between Monaco (100%) and the Kosovo (20.5%) in Europe, and between Morocco (51.0%) and Western Sahara (0%) in Africa. In areas such as Western Sahara, where 522, 928 people live without any Internet access, it is impossible to operate a cloud computing business.

Table 4.6: Internet Penetration Statistics in Selected Countries (Source: Internet World Stats[919])

Country	Penetration(%population)	Highest	Lowest
Japan	79.5%		
North America	78.6%		
Europe	63.2%	Monaco 100%	Kosovo 20.5%
China	40.1%		
Total for Africa	15.6%	Morocco 51.0%	Western Sahara 0%
World total	34.3%		

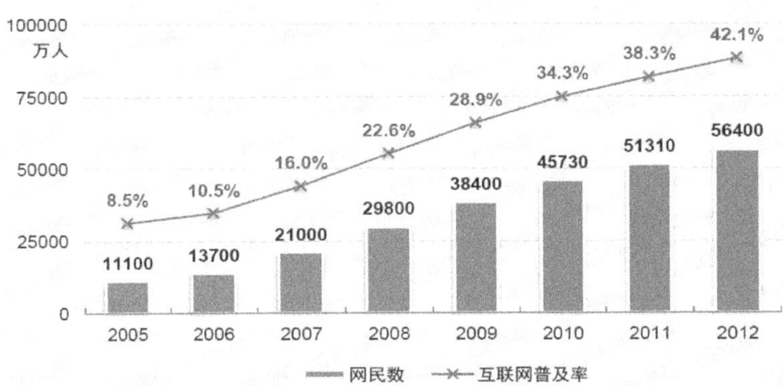

Fig. 4.3: Size of Internet Users and Internet Penetration Rate in China (Column: Number of Internet Users(10,000 persons); Line: Internet Penetration Rate) Source: CNNIC[920]

919 The statistics were updated for June 30, 2012, available at http://www.internetworldstats.com/ (last visited May 20, 2013).

In respect of China, although by the end of June, 2008, the amount of netizens had reached 253 million and leaped to first place in the world, the Internet penetration rate is relatively low (19.1%).[921] If the rate continues to increase at an average high rate of 4% per year, it will take another 15 years to achieve full Internet penetration. The construction of the Internet service infrastructure is a time consuming task.

4.4.5 Security Concerns for "Cloud"

Although cloud model provides great convenience for the usage of digital information, not everything is suitable for cloud computing. For example, in respect of some sensitive or privacy information, you might not want to post your data in the cloud but store locally. Big vendors, such as Amazon, Google, Microsoft, and Yahoo! have offered reliable services with strict privacy policies and strong security measures, thus some supporters argue that if you always encrypt your data before storing it on a vendor's cloud,[922] together with the vendor's security measures, your data may be much safer than if it were stored in-house.[923] However, "there are laws and policy that allow the government freer access to data on a cloud than on a private server. For example, the Stored Communications Act allows the FBI access to data without getting a warrant or the owner's consent."[924]

The security of users' private information is an issue noticeable in China. Due to national security considerations, several popular foreign video sharing websites or social networking services, such as YouTube, Facebook, and Twitter

---

920 China Internet Network Information Center (CNNIC), *The 31th Statistical Report on Internet Development in China*, January 2013, p.11, available at http://www.cnnic.cn/hlwfzyj/hlwxzbg/hlwtjbg/201301/P020130122600399530412.pdf (last visited May. 20, 2013).

921 See China Internet Network Information Center (CNNIC), *The 22nd Statistical Report on Internet Development in China*, Abridged Edition, July, 2008, available at http://www.cnnic.cn/download/2008/CNNIC22threport-en.pdf (last visited Apr. 20, 2011).

922 For example, you can use programs, such as open-source TrueCrypt (www.truecrypt.org) and PGP (www.pgp.com), to encrypt your files before you sending it to a cloud vendor so that only those with a password could access them.

923 See VELTE, VELTE & ELSENPETER, *supra* note 841, at 31.

924 *Id.* at 26.

are blocked in the PRC.[925] Similarly, Google complains of its service in China being blocked from time to time.[926] As VC Ramesh criticizes:

> Yet Google never materialized as a threat, for reasons that no one can agree on. What's clear is that Google's China service was mysteriously unreliable, particularly when the Chinese government was angry with the U.S. Google also documented several instances when it was the victim of what's called domain name system poisoning; users typed in Google.cn and found themselves at Baidu instead. Some Google executives privately say they think the Chinese government was never going to let Google succeed. MacKinnon, at the New America Foundation, says the opaque Chinese regulatory system makes it difficult to prove there was official discrimination. "Did Google face a very poisonous atmosphere in China? Sure," she says. "Whether this was the result of top down policy or the fact that Baidu's government relations were simply better is impossible to say."[927]

Admittedly, an evaluation of Chinese Internet censorship is not the subject of this paper. It seems certain, however, that the future development and prevalence of cloud computing will to a great extent depend on the existence of a healthy and orderly Internet environment which provides security for users' privacy and personal information. If there are no regulatory restrictions on unauthorized access by government agencies, users will hesitate to store their content in the cloud.[928]

4.4.6 The Dilemma Between Access and Control

As mentioned above, the physical precondition of cloud computing is access via the Internet. Access control TPMs have been widely employed in order to provide security to the information stored in clouds. The perfect control provided by TPMs might, however, generate negative effects on subsequent

---

925 See e.g., *YouTube Blocked in China*, CNN TECH, Mar. 25, 2009, available at http://articles.cnn.com/2009-03-25/tech/youtube.china_1_video-sharing-youtube-tibet-and-taiwan?_s=PM:TECH (last visited Apr. 11, 2011).

926 In response, Google moved China search service to Hong Kong attempting to strike a balance between resolving its dispute with Chinese censorship and keeping a foot in the world's largest Internet market. See Alexei Oreskovic, *Google Moves China Search Service to Hong Kong*, REUTERS, Mar. 22, 2010, available at http://www.reuters.com/article/2010/03/22/us-china-google-idUSTRE62L14V20100322 (last visited Apr. 11, 2011).

927 VC RAMESH, CHINA PLAY STOCKS, § Internet, Social Media and E-commerce.

928 For a further discussion of security issues by government use of cloud computing, see e.g., Wyld, *supra* note 876, at 6-12.

creativity. Certain infrastructure is thus needed in order to promote access to knowledge and creativity.

Admittedly, given the fact that cloud computing itself is a newly emerging business, exploration of the balance between access and control in the cloud model has just gotten under way. At present, there is no mature and successful experience to learn from. In respect of China, the cloud computing business is not well-developed at present, partly because of the low penetration rate of the Internet. Together with the limitations both in the copyright statute and in the implementation "dual system", it might be more difficult and time-consuming for China to establish the effective infrastructure necessary to promote access to knowledge and incentives to create at the same time.

4.5 Summary

In this chapter, I have compared the "TPMs & contract" model with the other three alternatives of digital content providing – the "levy system" model, the "device-centric" model, and the "cloud" model. The main difference between the "TPMs & contract" model and the other three alternatives is its opt-in or opt-out approach. In the "TPMs & contract" model, the default rule is that users can not use until they take the initiative to request permission to do so. In comparison, the default rule in other alternatives is that users can use unless they are definitely excluded from doing so by right holders, thus the onus is on right holders to take the initiative to provide options or conditions for users to choose or negotiate.

Arguably, each business model has its merits and flaws to some extent, especially in respect of private use. The prominent advantages of the "TPMs & contract" model are the low cost of anti-piracy and discriminatory pricing. However, it has limitations including the threat to consumers' computer safety, the weakening of users' enjoyment, and the prevention of sharing and communication. Therefore, it is necessary to allow for the deletion of TPMs in certain cases, such as for the sake of fair use, and to introduce a third-party institution to tackle the deletion and supervise the proper use of works after TPMs have been deleted.

The "levy system" model has been applauded for allowing free use in exchange for the payment of a levy. On the other hand, however, its non-discriminatory pricing, and especially the imposition of copyright levies on the sale of multi-purpose devices has been criticized for penalizing those who do not infringe copyright and encouraging copyright infringement amongst consumers. In response, a proper levy should be limited to equipment, devices or media whose primary purpose is to reproduce works. In addition, copyright societies are needed for collecting, managing and distributing the levies.

In the "device-centric" model, the attraction of the digital content is its cheap price, open source and the fact that it is TPMs-free. However, it is often subject to bundled sales with devices and a lack of place-shifting and harmonization. Consequently, preventing bundled sales with devices, and allowing place-shifting and harmonization are the required social conditions for this business model to be met.

The "cloud" model is regarded as the next paradigm shift in computer technology which has the ability to greatly facilitate users' enjoyment of digital works. As long as there is an Internet connection, users can access the content stored in the cloud at any time and at any place. The technology groundwork, therefore, demands a full penetration of the Internet. In addition, in order to secure the safety of the digital content stored in the cloud, TPMs on access control are required. This, however, also generates access control problems and negative effects on subsequent creation. Accordingly, certain infrastructure is needed in order to promote access to knowledge and creativity.

There are, however, many difficulties involved in reaching these social conditions in the Chinese context. For instance, in both the "TPMs & contract" model and the "levy system" model, lawful private use should be protected and illegal piracy punished, but it is very difficult to separate the two in practice. The management institutions needed to do so might also be subject to agency capture and corruption. In the case of the "device-centric" model, in the environment of the current unsound market economy in China, a firm might violate the principle of free competition for short-term commercial profits. Moreover, as a developing country, it is very time consuming to reach a state of full Internet penetration. Together with the security concerns for the "cloud", and the difficulty of managing an infrastructure to balance access and control, there is a long way to go before the prevalence of cloud computing becomes a reality.

Chapter V Working Proposals of the Legal Treatment of Private Use in China: An Example of the "TPMs & Contract" Model

According to the institutional analysis approach to law, each institution has its merit and flaw to some extent. It is impossible to find the "best" institution but rather to search for the least imperfect alternative.[929] In deed, after reviewing the other possible alternatives of "TPMs and contract" model, especially the pros and cons of each business model for content provision, I find it becomes more and more difficult to say which one is the best or which one is better than another. Therefore, basically, I have no right answer concerning the choice of business models.[930] However, I do believe as long as we choose any model, certain social conditions should be met in order to be effective.

In this section, I intend to choose "TPMs & contract" model as an example to illustrate the necessary modifications both in law and social infrastructure. I select it because compared to other options, the overprotection of TPMs in China is a prominent problem desiderated to be solved. In contrast, since there is no levy system within the Chinese copyright law, "levy system" model demands to introduce copyright levy into the statute. In respect of "device-centric" model, the problem of bundling sale is less urgent because it could have recourse to the anti-competition law or consumer protection law. Concerning "cloud" model, its prevalence depends on high Internet penetration, which is quite time-consuming in China.

As I have explored in the previous chapter, from users' perspective, the shortcomings of "TPMs & contract" model include threatening computer safety, weakening enjoyment, and preventing sharing as well as communication.[931] Thus, in order to restore a delicate balance between stakeholders and overcome the drawbacks of a perfect control by TPMs, it is necessary to allow deleting TPMs in certain cases for the sake of fair use and other copyright limitations. Besides, third-party institution is needed to deal with the deletion and supervise

---

929 See KOMESAR, IMPERFECT ALTERNATIVES, *supra* note 51, at 3-13.

930 I agree with Professor Shin Teramoto's argument. He considers that: "'no right answer' is the 'right answer'. Because the technological and economical conditions are changing everyday, any answer we can reach 'today' is nothing more than an answer to the question under assumption of 'yesterday'."

931 See *supra* section 4.2.1.

the proper use of TPMs deleted works.[932] However, when reaching these social conditions in Chinese practice, flaws such as difficulty to distinguish between private use and piracy, as well as agency capture and corruption might occur on account of the realistic situations.[933] Consequently, I intend to provide the following working proposals of the legal treatment of private use in China in respect of "TPMs & contract" model.

5.1 Considerations for Determining Private Use

As mentioned above, private use is notoriously full of uncertainty both in theory and in application. Given the copying itself is physically the same, it is almost impossible to set an absolute clear boundary between private use and piracy, but depend on a context- originated determination.[934]

However, we should not give up doing the right thing because of difficulty. Private use serves important values of free speech, and provides underground nutrition for creativity by virtue of fostering the creativity-relevant human capital through private studying, reading, watching, listening and communication.[935] Just like the law forbids drunk driving, but not forbids drinking alcohol, consequently, copying in private use context should be allowed, while to forbid copying for public distribution, P2P file sharing, or other pirate activities.

Moreover, the presumption of innocence principle declares that "[e]veryone has the right to be presumed innocent, and treated as innocent, until and unless they are convicted according to law in the course of proceedings which meet the prescribed requirements of fairness."[936] Accordingly, private users have the right to be presumed innocent before the outcome of the trial. Law makers, public authorities, and copyright holders should not treat them as suspects of pirates and make artificial restrictions on their legitimate rights by virtue of either law or technological measures.

On account of the conventional criteria to separate private use from piracy – enjoyment and exploitation as well as the two sub-factors, including noncommercial use or commercial use, and within private domain or market

932 See *supra* section 4.3.1.

933 See *supra* sections 4.4.1 & 4.4.2.

934 See *supra* Introduction I.3 & 4, and section 3.1.

935 See *supra* sections 3.4 & 3.5.3.

936 HUMAN RIGHTS MONITORING: A FIELD MISSION MANUAL 192 (Anette Faye Jacobsen ed., The Netherlands: Martinus Nijhoff Publishers 2008).

place – are full of vagueness,[937] it is necessary to consider more detailed factors for determining private use. In my opinion, there are at least two considerations deserved into account.

5.1.1 Determining Noncommercial Use: Weighing the Impact of Particular Private Use on Copyright Incentives and Liberties

Given the judgment of private use is fully context-based, and there is more uncertainty than clarity around whether specific use is commercial or noncommercial,[938] it is impossible to draw a clear line between commercial use and noncommercial use.[939] Consequently, when determining whether a particular use is noncommercial private use or not, it is useful to make a quantitative analysis weighing between the effects of the use on copyright incentives and liberties.

According to incentive theory, copyrights are "understood as incentives to authors, without which they might not create intellectual works."[940] Copyright incentives promote freedom of expression by virtue of promoting the creation of new expressions.[941] However, "[a]t the same time, incentive economics teaches us: copyright carries with it some of the problems of monopoly pricing and deadweight loss. As a result, copyright should not be expanded beyond the minimum necessary to provide authors with an incentive to create and make their works available to the public."[942]

In this way, copyright "[p]rivileges are not unlimited and an owner has to take account of possible repercussions....The term 'liberty' may further explain this: the right to use is correlative to a negative proposition in respect of the duties of the right-holder."[943] Concerning copyright, individual users always

---

937 See *supra* Introduction I.3, and section 3.1.

938 See Creative Commons, *supra* note 434.

939 See *supra* section 3.1.1.3.

940 Michael D. Birnhack, *Copyrighting Speech: A Trans-Atlantic View*, in COPYRIGHT AND HUMAN RIGHTS: FREEDOM OF EXPRESSION – INTELLECTUAL PROPERTY – PRIVACY 40 (Paul L.C Torremans, ed., The Hague: Kluwer Law International 2004).

941 See William M. Landes & Richard A. Posner, *An Economic Analysis of Copyright Law*, 18 J. LEG. STUD. 325 (1989).

942 Netanel, *supra* note 677, at 17-18.

943 LIOR ZEMER, THE IDEA OF AUTHORSHIP IN COPYRIGHT 62 (England: Ashgate Publishing Limited 2007).

retain a liberty to use the copyrighted works as provided for in the doctrine of fair use and other copyright limitations.

Generally speaking, there are four different situations when the effects of a use on copyright incentives and liberties can be compared:[944]

Table 5.1: Effects of a Particular Use on Copyright Incentives and Liberties

Effects on Copyright Incentives	Effects on Copyright Liberties	Estimation
Harming	Not Enhancing	Infringement
Little Harming	Greatly Enhancing	Legal
Not Harming	Not Enhancing	Inessential
Harming	Enhancing	Careful Balancing

From the viewpoint of the above table, the most difficult circumstance to determine is when the particular use both enhances copyright liberties and undermines copyright incentives. A careful balancing is required in order to harmonize the conflict between protecting the author's interest and promoting the public interest. As *The Copyright Principles Project* points out:

> Maintaining a balance between a copyright owner's exclusive rights and the public's right to use such works free from copyright owner control is critical for a well-designed copyright law. Such a balance allows copyright law to provide the proper incentive to encourage authors and their financial backers to create and to distribute new works, while permitting users and follow-on creators to engage in activities that further knowledge and progress.[945]

The current regulations provide private use in an ambiguous manner, referring to the use of the published work of another for the user's own private study, research, or self entertainment.[946] Determining whether a use constitutes a private use therefore heavily depends on a subsequent "three-step test".[947] That is to say, the private use exemption must not conflict with the normal exploitation of the work, and must not unreasonably prejudice the legitimate rights and interests of the copyright owner.[948]

---

944 See Litman, *supra* note 4, at 1911-12.

945 Pamela Samuelson et al., *supra* note 171, at 33.

946 See Article 22 (1) of the 2010 Chinese Copyright Law.

947 See *supra* sections 1.2.2 & 1.2.3.

948 See Article 21 of the 2002 Chinese Copyright Law Implementing Regulations.

The difficulty lies in determining how to estimate the "normal exploitation of the work" and "legitimate rights and interests of the copyright owner". Generally speaking, there are two different types of claims of harm in intellectual property – actual harm and potential harm:

> Actual harm occurs when there is a reduction in the profits of the producer below their level prior to a new unauthorized use. Actual harm does not occur, however, if the unauthorized use leaves profits from all previous uses unaffected.... [P]otential harm ...occurs if the new use reduces profits below the level they would have reached had the producer been able to exploit the market served without authorization. In other words, potential harm arises where a proprietor could have and would have been able to supply the product and receive compensation. Unlike actual harm, potential harm occurs even if existing markets are unaffected.[949]

Copyright holders seek expansive protection of their works, covering not only actual or direct harm, but also potential or indirect harm. "Within the copyright community, many people argue that private use damages the copyright proprietor economically by displacing potential sales and thus profits."[950] For instance, "some copyright owners regard personal uses as implicating the same core reproduction right as commercial uses do. Insofar as there is a commercial demand for personal use copies, copyright owners may want compensation."[951] Accordingly, they argue that "adoption of a personal use exemption might well lead to higher prices for copyrighted goods to compensate rights holders for the extra copies purchasers may make, regardless of whether these purchasers actually make or value extra copies."[952]

In fact, claims of potential harm reflect the competing interest conflict between stakeholders, especially in respect of the new opportunities provided by technological development. For copyright holders, the new opportunities represent new and profitable markets; while for users, these opportunities provide enhanced flexibility, widespread availability and inexpensive access.[953] As the Office of Technology Assessment of the US Congress points out:

> The criterion of potential harm makes intellectual property rights considerably more powerful than that of actual harm. This added strength of rights under a theory of potential harm stems from the fact that the potential uses to which a work can be put are unlimited in number and variety. Once the criterion of potential harm is accepted,

---

949 US CONGRESS, *supra* note 534, at 199, citing Stanley M. Besen, *Economic Issues Relating to New Technologies and Intellectual Property*, Contract Report prepared for OTA by the Rand Corp., December 1984, p.46.

950 US CONGRESS, *supra* note 534, at 197.

951 Pamela Samuelson et al., *supra* note 171, at 54.

952 *Id.*

953 See US CONGRESS, *supra* note 534, at 200.

> its amount is limited in principle only by the activities covered by intellectual property rights. So long as a case can be made that a user *could have and would have* remunerated the proprietor for the use, all conceivable uses that infringe a right could be said to cause potential harm.[954]

Pursuant to the principle of legality, the categories and scope of intellectual property rights should be provided by law. Without authorization by law, any other institutions or individuals cannot create new rights by themselves.[955] Therefore, potential harm cannot itself justify the extension of a right.[956] Instead, we should "define commercial use narrowly enough to capture direct financial gain and exclude more indirect benefit".[957]

In a word, when determining noncommercial use, we need a sensitive and careful evaluation to weigh the impact of particular private use on copyright incentives and liberties. That is to say, we should be cautious about the tendency for outstretched explanations to expand the claims of harm to potential harm, or further replacing "unauthorized use" with "commercial use", and "noncommercial use" with "licensed use".[958]

5.1.2 Restrictive Interpretation on Private Domain

When determining whether a certain use is a private use, another requirement is that the use must be within the private domain. Given the fact that piracy refers to the distribution of copyrighted copies to more than a selected group of people, a minimum standard for private use is that it should be within a selected group of people. However, if we consider the possible negative effect on copyright incentives, I suggest adopting a restrictive interpretation of the private domain.

---

954 *Id.*

955 See Zheng Shengli, Zhishi chanquan fading zhuyi [*The Principle of Legality of Intellectual Property*], Zhongguo sifa wang [China Private Law Net], available at http://www.privatelaw.com.cn/new2004/shtml/20040527-191705.htm (last visited Mar. 10, 2011); Li Yang, Lüe lun zhishi chanquan fading yuanze [*Research on the Principle of Legality in Intellectual Property*], Zhongguo sifa wang [China Private Law Net], available at http://www.privatelaw.com.cn/new2004/shtml/20041004-090722.htm (last visited Mar. 10, 2011); Li Jianhua, Lun zhishi chanquan fading yuanze: jian lun woguo zhishi chanquan zhidu de chuangxin [*The Principle of Legality in Intellectual Property: Innovation of the Chinese Intellectual Property System*], Zhongguo minshang falü wang [China Civil and Commercial Law Net], available at http://www.civillaw.com.cn/qqf/weizhang.asp?id=36878 (last visited Mar. 10, 2011).

956 See US CONGRESS, *supra* note 534, at 200.

957 See Litman, *supra* note 4, at 1914.

958 See *supra* section 3.1.1.3.

That is to say, the scope of private use should be even narrower than the selected group of people, such as classmates, colleagues, wedding guests, or other ordinary social acquaintances. It should denote only family members and close friends. In this sense, private use occurs in a sphere which is out of the control of copyright holders' exclusive rights.[959]

This restrictive interpretation on private use is especially important in China in order to preserve the private use limitation in the digital copyright scheme. One important reason to exclude private use from the fair use exception to the information network transmission right and exceptions to the anticircumvention provisions of TPMs is due to the great pressure of the anti-piracy battle in the digital environment.[960]

The significance of determining private use is to safeguard that those who meet its requirements do not need to obtain authorization or, sometimes, to pay remuneration.[961] If we retain a broad perception of the private domain and generally regard all uses among a selected group of people – such as copying and transmiting copies of copyrighted works among a large number of classmates or company employees – as constituting private use, it might greatly undermine the legitimate interests of right holders and harm their incentives to create.

Therefore, a restrictive interpretation of the private domain limited to only family members and close friends would be helpful in efforts to keep a delicate balance between copyright incentives and liberties. This balance could effectively promote copyright liberties without unreasonably prejudicing the legitimate rights and interests of the copyright owners.

5.2 Legal Protection of TPMs Should Follow Copyright Limitations

According to the WCT and the WPPT, protected technical measures must be employed to protect uses by right holders in connection with the exercise of their rights in respect of their works, which are not authorized by the copyright holders or permitted by law.[962] In other words, the legal protection of TPMs must not affect copyright limitations.

959 See *supra* section 3.1.2.
960 See *supra* sections 1.3 & 2.1.
961 In general, private copying would involve a copyright levy. Other private uses basically belong to fair use and are free of charge.
962 See Article 11 of the WCT, and Article 18 of the WPPT.

As a member state of the WCT and the WPPT, when implementing the obligations concerning technological measures, China should also follow the prerequisite provision on TPMs protection.[963] However, as an important copyright limitation, private use has been excluded from the scope of both the fair use limitation on the information network transmission right, and exceptions to the anticircumvention provision of TPMs.[964] That is to say, the current legal framework of TPMs does not sufficiently obey the basic precondition provided by the WIPO Internet Treaties.

Furthermore, abolishing the private use limitation of digital copyright as an anti-piracy measure is not a reasonable choice as it fails to catch the deep-seated causes of the piracy dilemma in China.[965] Given the fact that private use is a significant copyright liberty, and promotes subsequent creativities from a dynamic efficiency perspective, it is necessary to preserve it in the digital environment.[966]

Thus, in order to overcome the drawbacks of an overprotection for TPMs, and restore a delicate balance between stakeholders, the deletion of TPMs should be allowed in certain cases for the sake of private use or other copyright limitations. In the following section, I intend to provide some recommendations regarding efforts in both legal revision and judicial treatment.

5.2.1 Adding Private Use Limitation in Digital Copyright Framework

The current digital copyright framework provides copyright limitations in an explicitly listed way rather than as an open-ended standard. Article 6 of the 2006 Transmission Regulation enumerates eight fair use exceptions to the information network transmission right, none of them regarding private use. In addition, Article 12 of the 2006 Transmission Regulation provides four exceptions to the anticircumvention rules, including the nonprofit educational institution or scientific research exception, nonprofit uses for the benefit of the blind, administrative or judicial activities, and security testing.

However, the 2006 Transmission Regulation fails to clarify the relationship between TPMs and fair use. Outside of these four narrow enumerated exceptions to the anticircumvention provision, there is no acknowledgement of fair use

963 See *supra* section 3.2.1.

964 See *supra* sections 1.3.3 & 1.3.4.

965 See *supra* chapter II.

966 See *supra* sections 3.4 & 3.5.3.

exemption in general.[967] Private use has been excluded from the limitations of digital copyright. In this sense, as long as the copyright holders employ TPMs in their digital copyrighted works, private users are not allowed to circumvent the blockage even for legitimate fair uses acknowledged in the positive law. That is to say, copyright holders can legally employ TPMs to deprive private use limitation and other unlisted fair use.

In order to make the deletion of TPMs possible in certain cases for the sake of private use or other copyright limitations, it is essential to include a general principle in the statute to the effect that the TPMs regulations do not affect fair use. As an equitable policy tool, "[t]he purpose of fair use is twofold: to protect the copyright holder's market monopoly while preventing the market monopoly from being used to inhibit (rather than promote) learning."[968]

Notably, as a balancing policy tool, it is better to offer an open-ended standard rather than a closed enumeration.[969] It would be harmful to freeze the fair doctrine in the statute, especially during a period of rapid technological change.[970] The statute could of course include certain examples of fair uses.

---

967 See Feng Xiaoqing, *supra* note 200, at 37 (stating that the four fair use exemptions to the anticircumvention provision of TPMs in the 2006 Transmission Regulation are too narrow and need to be improved).

968 Patterson, *supra* note 30, at 274.

969 See Michael J. Madison, *Fair Use and Social Practices*, in INTELLECTUAL PROPERTY AND INFORMATION WEALTH: ISSUES AND PRACTICES IN THE DIGITAL AGE, VOL. 1, COPYRIGHT AND RELATED RIGHTS, 184 (Peter K. Yu ed., Connecticut: Praeger Publishers 2007) (indicating that pursuant to the statements of the legislative history to section 107, the "equitable" nature of the fair use doctrine renders it incapable of precise definition except by reference to the four factors stated in the statute); a similar observation could be found in the US House Report, see H. R. Rep. No. 94-1476, at 65 (1976) ("Indeed, since the doctrine is an equitable rule of reason, no generally applicable definition is possible, and each case raising the question must be decided on its own facts. On the other hand, the courts have evolved a set of criteria which, though in no case definitive or determinative, provide some gauge for balancing the equities. These criteria have been stated in various ways, but essentially they can all be reduced to the four standards which have been adopted in section 107…").

970 See H. R. Rep. No. 94-1476, at 66 (1976) (The statement in the House Report suggests that "The bill endorses the purposes and general scope of the judicial doctrine of fair use, but there is no disposition to freeze the doctrine in the statute, especially during a period of rapid technological change. Beyond a very broad statutory explanation of what fair use is and some of the criteria applicable to it, the courts must be free to adapt the doctrine to particular situations on a case-by-case basis"); Madison, *id.* at 185 (when analyzing the four factors of section 107 of the US Copyright Law, the author indicates that "the four factors implement the doctrine, rather than define it, gives courts a contextual standard by which application of the four factors can be measured.").

These would not, however, inevitably constitute lawful fair uses, but rather be subject to a fair use test.[971] The application of fair use depends upon a case-by-case or work-by-work basis considering the kind of work being used and the kind of use one is making of the work.[972]

Regarding private use, there are two considerations: first, admitting a private use limitation on the information network transmission right. As long as a particular use satisfies the aforementioned noncommercial requirement of private use, and is limited to family members and close friends, it belongs to the fair use exception to the information network transmission right.[973] In other words, whether the claim of a private use limitation is valid or not depends on a case-by-case fair use analysis. The law should not exclude the application of a private use limitation *ex ante* by virtue of ignoring it in the enumeration list.[974]

Second is allowing for the circumvention of TPMs in the case of private use. That is to say, a copyright user for the sake of private use may circumvent the technical measures without commiting a copyright infringement. Meanwhile, copyright holders should offer necessary convenience for the users' application, such as by giving them access passwords. Of course, private users need to bear the cost of applying for the deletion of TPMs from a third-party institution will be discussed later, and also ensure the proper use of works after TPMs have been deleted.[975]

---

971 See S. Rep. No. 94-473, at 62 (1975) (The statements contained in the Senate Report indicates: "In any event, whether a use referred to in the first sentence of section 107 is a fair use in a particular case will depend upon the application of the determinative factors, including those mentioned in the second sentence.")

972 See *Campbell v. Acuff-Rose Music, Inc.*, 510 U.S. 569, 577, 29 U.S.P.Q. 2d (BNA) 1961, 1965 (1994) (indicating that the fair use doctrine "calls for case-by-case analysis"); Patterson, *supra* note 30, at 276.

973 A project entitled *The Copyright Principles Project: Directions for Reform* proposed by 20 leading US scholars, suggests the directions for reform. Although there is no consensus on the private use issue, the project suggests that private, non-commercial use that does not compete with or otherwise undermine the author's market should be protected as fair use. See Pamela Samuelson et al., *supra* note 171.

974 *Cf.* Professor Zhang Jin indicates that the fundamental issue of private copying is whether it is fair use. China should revise its copyright law in order to respond to the challenges of private copying by digital technology. See Zhang Jin, Banquan fa zhong siren fuzhi wenti yanjiu – cong yinshuaji dao hulianwang [RESEARCH ON PRIVATE COPYING IN COPYRIGHT LAW – FROM THE PRINTING PRESS TO THE INTERNET] (Beijing: Zhongguo zhengfa daxue chubanshe [China University of Political Science and Law Press] 2009).

975 For instance, private user must not provide others with the technologies, apparatus or components used for circumventing technological measures, and must not infringe upon

5.2.2 A Restrictive Reading of "Effectiveness" in Anticircumvention Rules

The core issue in anticircumvention rules is the definition of technological protection measures. Here is the definition provided in the Chinese positive law:

> Technical measures mean the *effective* technology, apparatus or components used for preventing from or restricting any glancing through, admiring the works, performance, audio and video products without the right holder's permission or providing the public with the works, performance, audio and video products via network transmission.[976]

Notably, the Chinese copyright law does not distinguish between the access control measures and copyright protection measures, nor does it differentiate the protection standards just as the DMCA does.[977] Article 4 of the 2006 Transmission Regulation states that in addition to the making or selling of devices or services that are used to circumvent, the act of circumvention itself is also prohibited. Articles 18 and 19 of the 2006 Transmission Regulation further provide civil, administrative, and criminal liability for the circumvention of TPMs. Consequently, any circumvention except where otherwise provided in laws or administrative regulations constitutes copyright infringement.

It can thus be seen that the anticircumvention rules in the Chinese statute are quite broad. In order to retain a delicate balance among stakeholders, and minimize the potential negative effect on copyright liberties, and especially to prevent the overprotection of TPMs, it is necessary to adopt a restrained attitude towards the application of anticircumvention rules in practice.

Given that the word "effective" is the keyword in the abovementioned TPMs definition, and that there is no explicit explanation regarding what that word means, the restraint on the application of anticircumvention rules will mainly depend on a restrictive interpretation of their "effectiveness".

Actually, it is not an easy task to locate a clear notion of "effectiveness". For instance, the WCT and the WPPT indicate that the protected TPMs shall be "effective" and "in connection with the exercise of" copyrights.[978] However, the Treaties offer no definitions whatsoever. Of course, effectiveness cannot mean

the legitimate rights and interests enjoyed by copyright holders. See Article 12 of the 2006 Transmission Regulation.

976 Article 26 of the 2006 Transmission Regulation.

977 See *supra* section 3.2.2.

978 See Article 11 of the WCT, and Article 18 of the WPPT.

impervious or undefeatable.[979] Instead, perhaps "[t]he ability of the average user having average skills to defeat the means without special assistance (and thereby frustrate rights-holders' intent) should serves as a yardstick when measuring the effect."[980]

"Hence, according to the restrictive interpretation, a TPM is not 'effective' either when it is defective or when the disruption it generates is unrelated to the protection of copyrights."[981] This narrow reading of "effectiveness" includes two criteria: first, a TPM should be reliable. A TPM is not effective when it does not function properly. Besides, a TPM is not effective when it interferes with the normal functioning of the attached equipment or service. For instance, if a copy-control measure interferes with the playability of programs on a TV set or VCR, the measure cannot be regarded as effective.[982]

Second, a TPM is effective when it acts in relation to its aim, which is to protect the exercise of copyright. A TPM is not effective when either it is unrelated to copyright protection, or it exceeds the scope of copyright, such as being added on works in the public domain.

In a word, the notion of "effectiveness" should be understood as the capacity of the technology, apparatus or components to reach the goal of exercising a right holder's copyright to control the exploitation of copyrighted works, without preventing the normal functioning of the attached equipment or service. "Of course, the notion of effectiveness will in the end remain a matter of judicial interpretation, but this definition should help courts go in the right direction."[983]

More importantly, this restrained interpretation approach could serve as a build-in balancing policy tool by virtue of considering "effectiveness" both as a factual inquiry and a normative standard, and adjusting the scope of copyright protection. As Zohar Efroni points out:

> Effectiveness may surely call for a technical inquiry testing technical vulnerability to attacks, but it may also call for a normative evaluation. Effectiveness, as shown, is a threshold condition, and holding a TPM ineffective would deny its eligibility for anticircumvention protections. As such, legislatures (and particularly judges) may use it as an instrument to monitor the breadth of anticircumvention laws. This

---

979 See RICKETSON & GINSBURG, *supra* note 478, § 15.11.

980 EFRONI, *supra* note 878, at 304.

981 *Id.*

982 See JORG REINBOTHE & SILKE VON LEWINSKI, THE WIPO TREATIES 1996, 145 (London: Butterworths Lexis Nexis 2002).

983 ESTELLE DERCLAYE, THE LEGAL PROTECTION OF DATABASES: A COMPARATIVE ANALYSIS 210 (UK: Edward Elgar 2008).

includes disqualifying TPMs when litigators misuse anticircumvention provisions for purposes inimical to copyright policy.[984]

5.3 Regularizing the Third-Party Institution Management Infrastructure

If adding a private use limitation to a digital copyright statute is successful, and a restrictive reading of "effectiveness" in the anticircumvention rules is adopted in practice, whenever conflicts between private use and TPMs protection occur, users may have recourse to the court. However, on account of high litigation costs including the cost of expensive expertise and time, perhaps we need an alternative dispute settlement mechanism with high convenience, efficiency, easy access and relatively low costs.

5.3.1 Choosing the NCAC as the Third-Party Institution

Regarding the question of institutional choice and design, there are mainly four alternatives: the copyright administrative agency, copyright tribunals, collective management organizations, and consumer organizations. Given the fact that collective societies and consumer organizations raise suspicions of bias, the intermediacy of an independent third party would obtain more public trust. The options there are to either build new administrative agencies or judicial bodies, or to use the existing ones to tackle conflicts between right holders and users relating to the application of TPMs.[985]

The subsequent question is which option would be more suitable in China, namely to set up new institution or to utilize an existing one? First of all, in the establishment of new institutions, the social costs are often high. Second, in China, administrative agencies together with the courts form the so-called copyright implementation "dual system", and there are already multi-headed copyright administrative agencies involved in copyright affairs.[986] The NCAC is the competent authority in charge of copyright, while other administrations including the Ministry of Culture, the SAIC, the Ministry of Commerce and the General Administration of Customs also participate in copyright protection issues. However, the multi-headed copyright administration induces overlap,

---

984 EFRONI, *supra* note 878, at 305.

985 See *supra* section 4.3.1.

986 See *supra* section 2.2.2.

inefficiency and inconsistency among agencies.[987] Therefore, if we set up a new third-party institution beyond the existing administrative and judicial "dual system" to tackle TPMs related conflicts, it might worsen the existing multi-headed administration situation. As an alternative, it would be a better choice to utilize the existing institutions.

In my opinion, the NCAC is the proper candidate. First, at present, administrative protection actually dominates the bipolar mechanism. The current copyright dual protection system possesses the characteristics of powerful administrative protection and correspondingly weak judicial protection.[988] Thus, the administrative process with the advantages of high expertise and flexibility might be more likely to achieve the objective of bringing complex phenomena under control than courts.[989]

Second, among the multi-headed copyright administrative agencies, the NCAC is the most appropriate option. As the copyright competent authority, the NCAC is in charge of copyright registration, administering statutory licenses, constituting copyright administrative regulations and investigating and punishing copyright infringement cases.[990] Therefore, the NCAC possesses more advantages than other administrative agencies in tackling the application of TPMs deletion and supervision of use thereafter.

5.3.2 Regularizing the NCAC's Management for Transparency

As mentioned above, the main potential problems with a third-party institution management infrastructure are agency capture and corruption.[991] According to Neil Komesar's theory on minoritarian bias, the concentrated high-stake interests of copyright holders often prevail over the dispersed small-stake interests of users in the legislative process as well as in public agency decision-making.[992] Thus, the NCAC might also be captured by copyright interest groups.

Moreover, the situation might be more complicated within the administrative dominant "dual system" framework. The rival powers of the

987 See *supra* section 2.2.2.2.1.
988 See *supra* section 2.2.2.1.
989 See Merrill, *supra* note 52, at 1049.
990 See *supra* section 2.2.2.1.1.
991 See *supra* section 4.4.2.
992 See KOMESAR, IMPERFECT ALTERNATIVES, *supra* note 51, at 76. For more discussion about the interest contest between right holders and users, as well as the minoritarian bias phenomenon, see *supra* section 3.5.2.2.2.1.

multi-headed administrative agencies either stagnate and generate policy inconsistencies, or exhibit "minoritarian bias" when one agency captures the legislature.[993] The limitations of the copyright "dual system" would to some extent affect the effective functioning of the NCAC to tackle TPMs related copyright disputes.

It is not the focus of this book to provide a fundamental solution to the limitations of the administrative dominant "dual system" and multi-headed copyright administration. Instead, I pay close attention to the transparency of the NCAC's settlement procedure in order to overcome the potential flaws of agency capture and corruption.

First, the NCAC should be empowered to rule on any conflict between copyright limitations and TPMs in order to make sure that copyright limitations are respected in applying the TPMs for copyright protection.[994]

Second, after receiving the claim, the NCAC should have the authority to investigate and make determinations on the important issues, such as the "effectiveness" of target TPMs, the noncommercial private use character of the use, and the question of whether the application of TPMs has deprived the user of a fair private use.

Third, upon request, the NCAC may order copyright holders who employ "effective" TPMs on their works to provide available means for users to exert copyright limitations. In case right holders do not follow the order within a period of time, for example 1 month, either the NCAC may issue an administrative injunction, or users may circumvent the TPMs without liability for violating anticircumvention rules. In order to reduce the cost of TPMs deletion in cases involving the exercise of copyright limitations, a mandatory or a strongly recommended registration of online copyrights may be adopted in order to backup TPMs information.[995] Of course, the decision of the NCAC is subject to the final ruling of justice.

---

993 See *supra* sections 2.2.2.2.1 & 2.2.2.2.2.

994 There are four branches of the NCAC, including the Administrative Department, Copyright Department, International Cooperation Department, and the Enforcement Department. Among those, the Enforcement Department is responsible for the supervision of digital network copyright affairs. Therefore, it might be also authorized to tackle TPMs related copyright disputes. For the functions of NCAC branches, available at http://www.ncac.gov.cn/cms/html/205/1800/List-1.html (last visited Feb. 2, 2011).

995 There are two alternative ways. First, one possible way is to adopt a mandatory registration of online copyrights which requires right holders to backup information of the adopted TPMs. By doing this, the cost of circumventing TPMs in case of copyright limitation can be reduced. This mandatory registration would be of great advantage to prevent the dead blockage of contents by TPMs when copyright holders cannot be

Last but not least, the NCAC should supervise the proper use of works after TPMs have been deleted. For instance, the NCAC may require the user to sign a guarantee in exchange for TPMs deletion to promise that none of the subsequent uses will exceed the scope of private use. In case users violate the guarantee, the NCAC may issue an injunction upon the request of right holders.

5.4 Summary

In this chapter, the "TPMs & contract" model was used as an example to illustrate the necessary modifications both in law and social infrastructure to provide a proper legal treatment of private use in China. The working proposal I suggest includes three aspects:

First, given the fact that the conventional criteria used to separate private use from piracy are quite vague, it is necessary to consider more detailed factors. On the one hand, when determining whether a particular use is noncommercial private use or not, we need a sensitive and careful evaluation to weigh the impact of a particular use on copyright incentives and liberties, and and to be vigilant against the expansion of claims to include those which claim potential harm. On the other hand, in order to retain a delicate balance between copyright incentives and liberties, I suggest a restrictive interpretation of private domain limited to family members and close friends.

Second, legal protection of TPMs should follow copyright limitations. In order to make the deletion of TPMs possible in certain cases for the sake of private use or other copyright limitations, it is essential to include a general principle in the statute to the effect that the TPMs regulations do not affect fair use. Concerning private use, it is necessary to admit a private use limitation on the information network transmission right, and allow for the circumvention of TPMs in the case of private use. Meanwhile, a restrictive reading of "effectiveness" should be adopted in applying anticircumvention rules, such that only a TPM which is reliable and for the sake of copyright protection should be regarded as effective.

Last but not least, the NCAC is the proper candidate to deal with the TPMs deletion and supervise the proper use of TPMs deleted works. In order to overcome the potential flaws of agency capture and corruption, the NCAC's settlement procedure should follow the transparency requirement.

---

found or die. Second, another option is to follow the registration of software copyright. Article 2 of Registration of Computer Software Copyright Procedures indicates that: "the copyright administrative department of the State encourages registration of software and gives priority protection to registered software." Similarly, the NCAC could encourage the registration of online copyrights and backup TPMs information.

Conclusion

With the use of works by individual users for their own study, research, or self-entertainment purposes, private use serves an important role not only relating to users' fundamental rights such as free speech and the right of autonomy and privacy, but also in contributing to the creative process by fostering creativity-relevant human capital as well as the ultimate objective of the Chinese copyright law in maintaining the public interest.

Private use is limited to the enjoyment of works rather than their exploitation, thus it is basically outside of the scope of copyright. Only when one wishes to make a copy does private use intersect with copyright. Given the fact that the act of copying itself is physically the same, the separation of private use from piracy fully depends on a context-based determination. In sum, private use refers to noncommercial use within the private domain, while piracy denotes commercial use in the market place. Furthermore, when determining whether a particular use is noncommercial private use or not, we need a sensitive and careful evaluation to weigh the impact of a particular use on copyright incentives and liberties, and adopt a restrictive interpretation of private domain limited to family members and close friends.

This book has critically evaluated the legal treatment of private use under Chinese digital copyright law, which provides overprotection for TPMs and abandons the private use limitation on digital copyright. Such a treatment in fact serves as an anti-piracy measure. However, it confuses private use with piracy, and fails to catch the deep-seated causes of the piracy dilemma in China, which are to be found in the law making process, the implementation of the law, adherence to the law and also in the market.

Instead, the legal protection of TPMs should follow copyright limitations. In order to do so, on the one hand, the statute must admit a private use limitation on the information network transmission right, and allow for the circumvention of TPMs in the case of private use. On the other hand, a restrictive reading of "effectiveness" should be adopted in applying the anticircumvention rules. In practice, the NCAC should be entitled to deal with the TPMs deletion and supervise the proper use of works thereafter.

This working proposal concerning the legal treatment of private use in the Chinese digital copyright regime does not serve as the sole answer. In addition to the "TPMs & contract" approach, there are at least three other alternatives for digital content provision – the "levy system", the "device-centric", and the "cloud". According to the institutional analysis approach to law, each institution has its merits and demerits to some extent. It is impossible to find the "best" institution, rather the search is for the least imperfect alternative.[996] Consequently, this paper is not about which business model is the best or which one is better than another, rather it has focused on exploring what kind of social conditions should be met in order for them to be effective.

Despite the fact that this paper offers no right answer concerning the choice of business models,[997] however, I do believe that the normative investigations on the fundamental issues about the objective of copyright law, and the interrelationship between copyright holders and users in this book might provide some hints for a comprehensive understanding concerning the right place of private use in the copyright regime, now or in the future.

---

996 See KOMESAR, IMPERFECT ALTERNATIVES, *supra* note 51, at 3-13.

997 It should be the copyright industries themselves that choose their own business model. See Pamela Samuelson et al., *supra* note 171, at 54-56 ("all industries are in the process of adjusting their business models, adjusting their distribution models, and trying to figure out how to respond to a dramatic change in how people acquire and use copyrighted works.").

Bibliography

Abraham, H.J. & Perry, B.A., Freedom and the Court: Civil rights and Liberties in the United States (NY: Oxford University Press 1998).

Adams, John, *Originality in Copyright: A Solution to the Database Problem?* in COPYRIGHT LAW: A HANDBOOK OF CONTEMPORARY RESEARCH (Paul Torremans ed., UK: Edward Elgar Publishing Limited 2007).

Aghion, Phillipe, García-Peñalosa, Cecilia & Howitt, Peter, *Knowledge and Development: A Schumpeterian Approach*, in ECONOMIC GROWTH AND MACROECONOMIC DYNAMICS: RECENT DEVELOPMENTS IN ECONOMIC THEORY (Steve Dowrick, Rohan Pitchford & Stephen J. Turnovsky eds., NY: Cambridge University Press 2004).

ALFORD, WILLIAM P., TO STEAL A BOOK IS AN ELEGANT OFFENSE: INTELLECTUAL PROPERTY LAW IN CHINESE CIVILIZATION (California: Stanford University Press 1995).

Alford, William P., *Law, Law, What Law? Why Western Scholars of China Have Not Had More to Say about Its Law*, in THE LIMITS OF THE RULE OF LAW IN CHINA (Karen G. Turner, James V. Feinerman, & R. Kent Guy eds., University of Washington Press 2000).

An Shoulian, Zhishi chanquan haishi sixiang kongzhi: dui zhongguo gudai fa de wenhua toushi [*Intellectual Proptery or Thought Control: A Cultural Interpretation of Chinese Acient Law*] in Falü de wenhua jieshi [A CULTURAL INTERPRETATION OF LAW] (Liang Zhiping ed., Shanghai: San Lian shudian [San Lian Bookstore] 1998).

ANDERMAN, STEVEN D. ED., THE INTERFACE BETWEEN INTELLECTUAL PROPERTY RIGHTS AND COMPETITION POLICY (Cambridge: Cambridge University Press 2007).

ANDERSEN, ROBIN & GRAY, JONATHAN, BATTLEGROUND THE MEDIA (Westport: Greenwood Press 2008).

APLIN, TANYA, COPYRIGHT LAW IN THE DIGITAL SOCIETY: THE CHALLENGES OF MULTIMEDIA (Oxford: Hart Publishing 2005).

ARMSTRONG, MICHAEL, A HANDBOOK OF HUMAN RESOURCE MANAGEMENT PRACTICE (10th ed., UK: Kogan Page Limited 2006).

Arrow, Kenneth J., *Economic Welfare and the Allocation of Resources for Inventions*, in THE RATE AND DIRECTION OF INVENTIVE ACTIVITY: ECONOMIC AND SOCIAL FACTORS, (Richard R. Nelson ed, Princeton: Princeton University Press 1962).

Ayres, Ian, *Playing Games with the Law*, 42 STAN. L. REV. 1291(1990).

AXELROD, ROBERT, THE EVOLUTION OF COOPERATION (NY: Basic Books 1984).

Bailey, Jane, *Deflating the Michelin Man: Protecting Users' Rights in the Canadian Copyright Reform Process*, in THE PUBLIC INTEREST: THE FUTURE OF CANADIAN COPYRIGHT LAW (Michael Geist ed., Toronto: Irwin Law 2005).

BAINBRIDGE, DAVID I., INTELLECTUAL PROPERTY (4th ed., London: Pitman Publishing 1999).

BAIRD, DOUGLAS, GERTNER, ROBERT & PICKER, RANDAL, GAME THEORY AND THE LAW (Cambridge: Harvard University Press 1994).

Bakardjieva, Antonina, *The WTO Dispute Settlement System and the Evolution of International IP Law: An Institutional Perspective*, in INTELLECTUAL PROPERTY RIGHTS IN A FAIR WORLD TRADE SYSTEM: PROPOSALS FOR REFORM OF TRIPS (Annette Kur & Marianne Levin eds., UK: Edward Elgar Publishing Limited 2011).

Balkin, Jack M., *Digital Speech and Democratic Culture: A Theory of Freedom of Expression for the Information Society*, 79 N.Y.U. L. REV. 1 (2004).

Balkin, Jack M., *Populism and Progressivism as Constitutional Categories*, 104 YALE L. J. 1935 (1995).

BARKAI, DAVID, PEER-TO-PEER COMPUTING: TECHNOLOGIES FOR SHARING AND COLLABORATING ON THE NET (Hillsboro, OR: Intel Corporation 2002).

Baron, Paula, *The Moebius Strip: Private Right and Public Use in Copyright Law*, 70 ALB. L. REV. 1227 (2007).

Basalisco, Bruno, Reid, Andy & Richards, Paul, *Interdependent Innovation in Telecommunications: Risk, Standardization and Regulation*, in REGULATION AND THE EVOLUTION OF THE GLOBAL TELECOMMUNICATIONS INDUSTRY (Anastassios Gentzoglanis & Anders Henten eds., UK: Edward Elgar Publishing Limited 2010).

Blanke, Jordan M., *Copyright Law in the Digital Age*, in SOCIAL, ETHICAL AND POLICY IMPLICATIONS OF INFORMATION TECHNOLOGY (Linda L. Brennan & Victoria Johnson eds., US: Information Science Publishing 2004).

Bechtold, Stefan, *Digital Rights Management in the United States and Europe*, 52 AM. J. COMP. L. 323 (2004).

Bechtold, Stefan, The *Present and Future of Digital Rights Management: Musings on Emerging Legal Problems*, in DIGITAL RIGHTS MANAGEMENT: TECHNOLOGICAL, ECONOMIC, LEGAL AND POLITICAL ASPECTS (Eberhard Becker et al. eds., Berlin: Springer 2003).

Becker, Lawrence C., *The Labor Theory of Property Acquisition*, 73 JOURNAL OF PHILOSOPHY 653 (1976).

Beebe, Barton, *An Empirical Study of U.S. Copyright Fair Use Opinions, 1978-2005*, 156 U. PA. L. REV. 549 (2008).

Bell, Tom, *Fair Use vs. Fared Use: The Impact of Automated Rights Management on Copyright's Fair Use Doctrine*, 76 N.C. L. REV. 557(1998).

BENKLER, YOCHAI, THE WEALTH OF NETWORKS: HOW SOCIAL PRODUCTION TRANSFORMS MARKETS AND FREEDOM (New Haven & London: Yale University Press 2006).

Benkler, Yochai, *A Political Economy of the Public Domain: Markets in Information Goods vs. The Marketplace of Ideas*, in EXPANDING THE BOUNDARIES OF INTELLECTUAL PROPERTY: INNOVATION POLICY FOR THE KNOWLEDGE SOCIETY (Rochelle Cooper Dreyfuss, Diane Leenheer Zimmerman & Harry First eds., Oxford University Press 2001).

Benkler, Yochai, *From Consumers to Users: Shifting the Deeper Structures of Regulation Toward Sustainable Commons and User Access*, 52 FED. COMM. L. J. 561 (2000).

BENTLY, LIONEL & SHERMAN, BRAD, INTELLECTUAL PROPERTY LAW (2nd ed., Oxford: Oxford University Press 2004).

BETTIG, RONALD V., COPYRIGHTING CULTURE: THE POLITICAL ECONOMY OF INTELLECTUAL PROPERTY (Colorado: Westview Press 1996).

Beyer, Lawrence Adam, *Intentionalism, Art, and the Suppression of Innovation: Film Colorization and the Philosophy of Moral Rights*, 82 NW. U. L. REV. 1011 (1988).

Birnhack, Michael D., *Copyrighting Speech: A Trans-Atlantic View*, in COPYRIGHT AND HUMAN RIGHTS: FREEDOM OF EXPRESSION – INTELLECTUAL PROPERTY – PRIVACY (Paul L.C Torremans, ed., The Hague: Kluwer Law International 2004).

Bloustein, Edward J., *Privacy as an Aspect of Human Dignity: An Answer to Dean Prosser*, 39 N.Y.U. L. REV. 962 (1964).

BOLLIER, DAVID, BRAND NAME BULLIES: THE QUEST TO OWN AND CONTROL CULTURE (NJ: John Wiley & Sons Inc. 2005).

Bollinger, Lee, *Book Review: Protect This Work of Expression: Clarifying the Unique Economics of Intellectual Property Rights, Steal This Idea: Intellectual Property Rights and the Corporate Confiscation of Creativity by Michael Perelman*, 44 SANTA CLARA L. REVIEW 1287 (2004).

BOSS, AMELIA H. & KILIAN, WOLFGANG, THE UNITED NATIONS CONVENTION ON THE USE OF ELECTRONIC COMMUNICATIONS IN INTERNATIONAL CONTRACTS: AN IN-DEPTH GUIDE AND SOURCEBOOK (Netherlands: Kluwer Law International 2008).

Bowrey, Kathy, *Who's Writing Copyright's History*, 18 EUROPEAN INTELLECTUAL PROPERTY REVIEW 322 (1996).

BOYLE, JAMES, THE PUBLIC DOMAIN: ENCLOSING THE COMMONS OF THE MIND (London: Yale University Press 2008).

BOYLE, JAMES, SHAMANS, SOFTWARE, AND SPLEENS: LAW AND THE CONSTRUCTION OF THE INFORMATION SOCIETY (Cambridge: Harvard University Press 1996).

Boyle, James, *The Second Enclosure Movement and the Construction of the Public Domain*, 66 SPG LAW & CONTEMP. PROBS. 33 (2003).

Brett, David H., *Digital Knowledge: Copyright Intellectual Property and the Internet*, in WORLD CONGRESS ON INTELLECTUAL CAPITAL READINGS (Nick Bontis ed., Boston: Butterworth-Heinemann 2002).

Brown, Brandon, *Fortifying the Safe Harbors: Reevaluating the DMCA in a Web 2.0 World*, 23 BERKELEY TECH. L.J. 437 (2008).

BUCKLE, STEPHEN, NATURAL LAW AND THE THEORY OF PROPERTY: GROTIUS TO HUME (Oxford: Oxford University Press 1991).

BUFORD, JOHN F., YU, HEATHER, & LUA, ENG KEONG, P2P: NETWORKING AND APPLICATIONS (US: Morgan Kaufmann Publishers 2009).

Bunker, Matthew D., *Eroding Fair Use: The "Transformative" Use Doctrine After Campbell*, 7 COMM. L. & POL'Y 1 (2002).

Burk, Dan L. & Cohen, Julie E., *Fair Use Infrastructure for Rights Management Systems*, 15 HARV. J.L. & TECH. 41 (2001).

BURRELL, ROBERT & COLEMAN, ALLISON, COPYRIGHT EXCEPTIONS: THE DIGITAL IMPACT (Cambridge: Cambridge University Press 2005).

Byron, Thomas M., *Up Feist's Loose Ends: A Probability Theory of Copyrightable Creativity*, 7 WAKE FOREST INTELL. PROP. L.J. 44 (2006).

Carrier, Michael, *Unraveling the Patent-Antitrust Paradox*, 150 U. PA. L. REV. 761 (2002).

Chen Huizhen & Xu Jun, Lun woguo zhishi chanquan liti shenpan moshi de goujian [*On the Construction of Chinese IP Solid Trials*], 4 Falü shiyong [JOURNAL OF LAW APPLICATION] 12 (2006).

COHEN, JULIE E. ET AL., COPYRIGHT IN A GLOBAL INFORMATION ECONOMY (NY: Aspen Publishers 2006).

Cohen, Julie E., *The Place of the User in Copyright Law*, 74 FORDHAM L. REV. 347 (2005).

Cohen, Julie E., *Copyright's Public-Private Distinction*, 55 CASE W. RES. L. REV. 963 (2005).

Cohen, Julie E., *DRM and Privacy*, 18 BERKELEY TECH. L. J. 575 (2003).

Cohen, Julie E., *A Right to Read Anonymously: A Closer Look at "Copyright Management" in Cyberspace*, 28 CONN. L. REV. 981 (1996).

COLSTON, CATHERINE & MIDDLETON, KIRSTY, MODERN INTELLECTUAL PROPERTY LAW (2nd ed., UK: Cavendish Publishing Limited 2005).

COOMBE, ROSEMARY J., THE CULTURAL LIFE OF INTELLECTUAL PROPERTIES: AUTHORSHIP, APPROPRIATION AND THE LAW (Durham: Duke University Press 1998).

COOMBS, ROD, SAVIOTTI, PAOLO, & WALSH, VIVIEN, ECONOMICS AND TECHNOLOGICAL CHANGE (New Jersey: Rowman & Littlefield 1987).

COOTER, ROBERT & ULEN, THOMAS, LAW & ECONOMICS (5th ed., Boston: Pearson Education 2008).

Cornish, William R., *Moral Rights under the 1988 Act*, 11 (12) EIPR 449 (1989).

Davies, Gillian, *The Convergence of Copyright and Author's Rights – Reality or Chimera?* 26 IIC 964 (1995).

DEAZLEY, RONAN, RETHINKING COPYRIGHT: HISTORY, THEORY, LANGUAGE, CHELTENHAM (UK: Edward Elgar 2006).

deBeer, Jeremy, *Constitutional Jurisdiction over Para-copyright Laws*, in IN THE PUBLIC INTERESTS: THE FUTURE OF CANADIAN COPYRIGHT LAW (Michael Geist ed., Toronto: Irwin Law 2005).

DEGRAFF, JEFF & LAWRENCE, KATHERINE A., CREATIVITY AT WORK: DEVELOPING THE RIGHT PRACTICES TO MAKE INNOVATION HAPPEN (U.S.: John Wiley & Sons, Inc., 2002).

DERCLAYE, ESTELLE, THE LEGAL PROTECTION OF DATABASES: A COMPARATIVE ANALYSIS (UK: Edward Elgar 2008).

Derclaye, Estelle, *Intellectual Property Rights and Human Rights: Coinciding and Cooperation*, in INTELLECTUAL PROPERTY AND HUMAN RIGHTS (Paul Torremans ed., Netherlands: Kluwer Law International 2008).

Díaz, Álvaro, *Free Trade Agreement and Intellectual Property: Impacts and Challenges*, in KNOWLEDGE GENERATION AND PROTECTION: INTELLECTUAL PROPERTY, INNOVATION AND ECONOMIC DEVELOPMENT (Jorge Mario Martínez-Piva ed., NY: Springer 2009).

Dietz, Adolf, *The Artist's Right of Integrity under Copyright Law: A Comparative Approach*, 25 IIC 177 (1994).

DIMITROV, MARTIN K., PIRACY AND THE STATE: THE POLITICS OF INTELLECTUAL PROPERTY RIGHTS IN CHINA (Cambridge University Press 2009).

Drahos, Peter, *The Regulation of Public Goods*, 7 J. INT. ECON. L. 321 (2004).

DRAHOS, PETER & BRAITHWAITE JOHN, INFORMATION FEUDALISM: WHO OWNS THE KNOWLEDGE ECONOMY? (NY: The New Press 2002),

DRAHOS, PETER, A PHILOSOPHY OF INTELLECTUAL PROPERTY (England: Ashgate Publishing Limited 1996).

Duan Ruilin, Zhishi chanquan fa gailun [AN INTRODUCTION TO INTELLECTUAL PROPERTY LAW] (Beijing: Guang ming ribao chubanshe [Guangming Daily Publishing House] 1988).

Durie, Robyn, *Moral Rights and the English Business Community*, 2 ENT. L. REV. 40 (1991).

EAGLETON, TERRY, LITERARY THEORY: AN INTRODUCTION (2d ed., US: University of Minnesota 1996).

Eckersley, Peter, *Virtual Market for Virtual Goods: The Mirror Image of Digital Copyright?* 18 HARV. J. L. & TECH. 85 (2004).

EFRONI, ZOHAR, ACCESS-RIGHT: THE FUTURE OF DIGITAL COPYRIGHT LAW (NY: Oxford University Press 2011).

Engelbrert, Antonina B., *Copyright from an Institutional Perspective: Actors, Interests and the Logic of Participation*, 4 REVIEW OF ECONOMIC RESEARCH ON COPYRIGHT ISSUES 65 (2007).

ERHARD, LUDWIG, PROSPERITY THROUGH COMPETITION (NY: Frederick A. Praeger, Inc. 1958).

FEATHER, JOHN, PUBLISHING, PIRACY, AND POLITICS: AN HISTORICAL STUDY OF COPYRIGHT IN BRITAIN (NY: Mansell 1994).

Fei Zhengqing, Weida de Zhongguo geming [THE GREAT CHINESE REVOLUTION 1800-1985] (Beijing: Shijie zhishi chubanshe [World Affairs Press] 2000).

Feng Xiaoqing, Woguo zhuzuoquan fa dui zuopin quanyi fenpei de junheng sixiang tanxi [*Research on the Balance of Interest Consideration in the Chinese Copyright Law*], 4 Zhishi chanquan [INTELLECTUAL PROPERTY]16 (1996).

Feng Xiaoqing, Shi lun zuozhe zai zhuzuoquan zhong de falü diwei [*The Legal Status of the Author in the Copyright System*], 4 Zhishi chanquan [INTELLECTUAL PROPERTY] 16 (1995).

Feng Xiaoqing, Jishu cuoshi yu zhuzuoquan baohu tantao [*Research on Technological Protection Measures and Copyright Protection*], 4 Faxue zazhi [LAW SCIENCE MAGAZINE] 20 (2007).

Feng Xiaoqing & Liu Shuhua, Shi lun zhishi chanquan de siquan shuxing jiqi gongquan hua quxiang [*On the Private Right Nature of Intellectual Property and Its Public Trend*], 1 Zhongguo faxue [CHINESE LEGAL SCIENCE] 61 (2004).

FICSOR, MIHÁLY, THE LAW OF COPYRIGHT AND THE INTERNET: THE 1996 WIPO TREATIES, THEIR INTERPRETATION AND IMPLEMENTATION (Oxford: Oxford University Press 2002).

FINNIS, JOHN, NATURAL LAW AND NATURAL RIGHTS (Oxford: Oxford University Press 1980).

Fisch, Alan, *Compulsory Licensing of Blacked-Out Professional Team Sporting Event Telecasts (PTSETS): Using Copyright Law to Mitigate Monopolistic Behavior*, 32 HARV. J. ON LEGIS. 403 (1995).

Fisher, Franklin M., *Detecting Market Power*, in ABA SECTION OF ANTITRUST LAW, ISSUES IN COMPETITION LAW AND POLICY (vol.1, US: American Bar Association 2008).

FISHER III, WILLIAM W., PROMISES TO KEEP: TECHNOLOGY, LAW, AND THE FUTURE OF ENTERTAINMENT (California: Stanford University Press 2004).

Foucault, Michel, *What Is an Author*, in TEXTUAL STRATEGIES: PERSPECTIVES IN POST-STRUCTURALIST CRITICISM (Josue Harari ed., Ithaca: Cornell University Press 1979).

Frischmann, Brett & Moylan, Dan, *The Evolving Common Law Doctrine of Copyright Misuse: A Unified Theory and Its Application to Software*, 15 BERKELEY TECH. L.J. 865 (2000).

Froomkin, Michael A., *The Death of Privacy?*, 52 STAN. L. REV. 1461(2000).

GANEA, PETER, PATTLOCH, THOMAS & HEATH, CHRISTOPHER EDS., INTELLECTUAL PROPERTY LAW IN CHINA (Hague: Kluwer Law International 2005).

Geller, Paul Edward, *Beyond the Copyright Crisis: Principles for Change*, 55 J. COPYRIGHT SOC'Y U.S.A. 165 (2008).

GEORGE, ROBERT P., ED. NATURAL LAW THEORY: CONTEMPORARY ESSAYS (Oxford: Clarendon Press 1992).

Gervais, Daniel J., *Intellectual Property and Human Rights: Learning to Live Together*, in INTELLECTUAL PROPERTY AND HUMAN RIGHTS (Paul Torremans ed., Netherlands: Kluwer Law International 2008).

GERVAIS, DANIEL J. ED., INTELLECTUAL PROPERTY, TRADE AND DEVELOPMENT: STRATEGIES TO OPTIMIZE ECONOMIC DEVELOPMENT IN A TRIPS-PLUS ERA (Oxford: Oxford University Press 2007).

Gervais, J. Daniel, *A Canadian Copyright Narrative*, in COPYRIGHT LAW: A HANDBOOK OF CONTEMPORARY RESEARCH (Paul Torremans ed., UK: Edward Elgar Publishing Limited 2007).

Gervais, Daniel J., *The Price of Social Norms: Towards a Liability Regime for File-Sharing*, 12 J. INTELL. PROP. L. 39 (2004).

Getz, Isaac & Lubart, Todd, *Creativity and Economics: Current perspectives*, in THE ROUTLEDGE COMPANION TO CREATIVITY (Tudor Rickards, Mark A. Runco & Susan Moger eds., Oxon: Routledge 2009).

GHIDINI, GUSTAVO, INTELLECTUAL PROPERTY AND COMPETITION LAW: THE INNOVATION NEXUS (UK: Edward Elgar Publishing Limited 2006).

Ghosh, Shubha, *Recognizing Authority in the Marketplace: The Curious and Ubiquitous Problems of Gray Markets*, in INTELLECTUAL PROPERTY AND INFORMATION WEALTH: ISSUES AND PRACTICES IN THE DIGITAL AGE, VOL.4, INTERNATIONAL INTELLECTUAL PROPERTY LAW AND POLICY (Peter K. Yu ed., Westport: Praeger Publishers 2007).

Ghosh, Shubha, *Pills, Patents, and Power: State Creation of Gray Markets as a Limit on Patent Rights*, 53 FLA. L. REV. 789 (2001).

Gimbel, Mark, *Some Thoughts on the Implications of Trusted Systems for Intellectual Property Law*, 50 STAN. L. REV. 1671 (1998).

Ginsburg, Jane C., *From Having Copies to Experiencing Works: The Development of an Access Right in U.S. Copyright Law*, 50 J. COPYRIGHT SOC'Y U.S.A. 113 (2003).

Ginsburg, Jane C., *Copyright Use and Excuse on the Internet*, 24 COLUM.VLA J.L. & ARTS 1 (2000).

Ginsburg, Jane C., *Authors and Users in Copyright*, 45 J. COPYRIGHT SOC'Y U.S.A. 1 (1997).

Ginsburg, Jane C., *A Tale of Two Copyrights: Literary Property in Revolutionary France and America*, 64 TUL. L. REV. 991 (1990).

Ginsburg, Jane C., *Moral Rights in a Common Law System*, 1 ENT. L. REV. 121 (U.K.) (1990).

GOLDSTEIN, PAUL, COPYRIGHT'S HIGHWAY: FROM GUTENBERG TO THE CELESTIAL JUKEBOX (revised edition, California: Stanford University Press 2003).

Goldstein, Paul, *Fair Use in a Changing World*, 50 J. COPYRIGHT SOC'Y U.S.A. 133 (2003).

Goldstein, Paul, *Copyright and Its Substitutes*, 45 J. COPYRIGHT SO'Y U.S.A. 151 (1997).

Goldstein, Paul, *Copyright in the New Information Age*, 40 CATH. U. L. REV. 829 (1991).

Goldstein, Paul, *Derivative Rights and Derivative Works in Copyright*, 30 J. COPYRIGHT SOC'Y U.S.A. 209 (1983).

Gordon, Wendy J., *A Property Right in Self-Expression: Equality and Individualism in the Nature Law of Intellectual Property*, 102 YALE L.J. 1533 (1993)

Gordon, Wendy J., *An Inquiry into the Merits of Copyright: The Challenges of Consistency, Consent, and Encouragement Theory*, 41 STAN. L. REV. 1343 (1989).

Gordon, Wendy, *Fair Use as Market Failure: A Structural and Economic Analysis of the Betamax Case and Its Predecessors*, 82 COLUM. L. REV. 1600 (1982).

GRIFFITHS, JONATHAN & SUTHERSANEN, UMA EDS., COPYRIGHT AND FREE SPEECH: COMPARATIVE AND INTERNATIONAL ANALYSES (Oxford: Oxford University Press 2005).

Gu Peidong, Ye lun Zhongguo faxue xiang hechu qu [*Where Will China's Legal Science Go?*], 1 Zhongguo faxue [CHINA LEGAL SCIENCE] 5 (2009).

Guibault, Lucie, *The Limit of Balancing Interests Through Copyright Levies*, in ETHICS AND LAW OF INTELLECTUAL PROPERTY: CURRENT PROBLEMS IN POLITICS, SCIENCE AND TECHNOLOGY (Christian Lenk, Nils Hoppe & Roberto Andorno eds., England: Ashgate Publishing Limited 2007).

GUIBAULT, LUCIE M.C.R., COPYRIGHT LIMITATIONS AND CONTRACTS: AN ANALYSIS OF THE CONTRACTUAL OVERRIDABILITY OF LIMITATIONS ON COPYRIGHT (The Hague; London: Kluwer Law International 2002).

GUIBAULT, LUCIE & HUGENHOLTZ, P. BERNT, THE FUTURE OF THE PUBLIC DOMAIN: IDENTIFYING THE COMMONS IN INFORMATION LAW (Netherlands: Kluwer Law International 2006).

HALBERT, DEBORA J., INTELLECTUAL PROPERTY IN THE INFORMATION AGE: THE POLITICS OF EXPANDING OWNERSHIP RIGHTS (Westport: Quorum Books 1999).

Han Ning & Li Yuling, Cong zhuzuoquan quanli qiongjie kan zhishi chanquan de xianzhi [*Comment on the Limitations of IPRs from the Exhaustion of Copyright Perspective*], 10 (Mid.) Fazhi yu shehui [LEGAL SYSTEM AND SOCIETY] 348 (2008).

Han Yuxiong & Li Huaizu, Guanyu zhongguo zhishi chanquan baohu shuiping de dingliang fenxi [*Quantitative Analysis on Chinese IPRs Protection Level*], 3 Kexue xue yanjiu [STUDIES IN SCIENCE OF SCIENCE] 377 (2005).

Hardy, Trotter, *Contracts, Copyright and Preemption in a Digital World*, 1 RICH. J.L. & TECH. 2 (1995).

Harold J. Berman, Falü yu zongjiao [LAW AND RELIGION] (Liang Zhiping trans., Beijing: Shangwu yinshuguan [The Commercial Press] 2002).

HAYNES, RICHARD, MEDIA RIGHTS AND INTELLECTUAL PROPERTY (UK: Edinburgh University Press 2005).

HEATH, CHRISTOPHER & SANDERS, ANSELM K. EDS., INTELLECTUAL PROPERTY IN THE DIGITAL AGE: CHALLENGES FOR ASIA (Hague: Kluwer Law International 2001).

HEATH, CHRISTOPHER ED., INTELLECTUAL PROPERTY LAW IN TAIWAN (Hague: Kluwer Law International 2003).

Helberger, Natali & Hugenholtz, P. Bernt, *No Place Like Home for Making a Copy: Private Copying in European Copyright Law and Consumer Law*, 22 BERKELEY TECH. L.J. 1061 (2007).

Herzog, Shai, Shenker, Scott & Estrin, Deborah, *Sharing Multicast Costs*, in INTERNET ECONOMICS (Lee W. McKnight & Joseph P. Bailey eds., Cambridge, Mass.: MIT Press 1997)

HOLTZMAN, DAVID H., PRIVACY LOST: HOW TECHNOLOGY IS ENDANGERING YOUR PRIVACY (CA: Jossey-Bass 2006).

Huang Renyu [Ray Huang], Zhongguo da lishi [CHINA: A MACRO HISTORY] (Beijing: San lian shudian [San Lian Bookstore] 1997).

Huang Shuhai et al, Guanyu woguo zhuzuoquan lifa de jidian sikao [*Considerations on Chinese Copyright Lawmaking*], in Banquan yanjiu wenxuan [SELECTED WORKS OF COPYRIGHT RESEARCH] (Zhongguo banquan yanjiuhui [Copyright Society of China] ed., Beijing: Shangwu yinshuguan [The Commercial Press] 1995).

Hugenholtz, P. Bernt, *Caching and Copyright: The Right of Temporary Copying*, 22 EUROPEAN INTELLECTUAL PROPERTY REVIEW 482 (2000).

Hughes, Justin, *The Philosophy of Intellectual Property*, 77 GEO. L. J. 287 (1988).

Jackson, Matt, *The Digital Millennium Copyright Act of 1998: A Proposed Amendment to Accommodate Free Speech*, 5 COMM. L. & POL'Y 61 (2000).

Jaszi, Peter, *On the Author Effect: Contemporary Copyright and Collective Creativity*, 10 CARDOZO ARTS & ENT. L.J. 293 (1992).

Jaszi, Peter, *Toward a Theory of Copyright: The Metamorphoses of "Authorship"*, 42 DUKE L. J. 455 (1991).

Ji Weidong, Fazhi zhixu de jiangou [CONSTRUCTION OF LEGAL ORDER] (Beijing: Zhongguo zhengfa daxue chubanshe [China University of Political Science and Law Press] 1999).

Jiang Ping, Jiang Ping wenji [SELECTED WORKS OF JIANG PING] (Beijing: Zhongguo fazhi chubanshe [China Legal Publishing House] 2000).

KALLAY, DINA, THE LAW AND ECONOMICS OF ANTITRUST AND INTELLECTUAL PROPERTY: AN AUSTRIAN APPROACH (UK: Edward Elgar 2004).

KAPLAN, BENJAMIN, AN UNHURRIED VIEW OF COPYRIGHT (NY and London: Columbia University Press 1967).

KASPER, WOLFGANG & STREIT, MANFRED E., INSTITUTIONAL ECONOMICS: SOCIAL ORDER AND PUBLIC POLICY (UK: Edward Elgar Publishing Limited 1998).

Karjala, Dennis S., *Copyright and Creativity*, 15 UCLA ENT. L. REV. 169(2008).

KEELING, DAVID T., INTELLECTUAL PROPERTY RIGHTS IN EU LAW (VOL. I): FREE MOVEMENT AND COMPETITION LAW (NY: Oxford University Press 2003).

KHAN, B. ZORINA, THE DEMOCRATIZATION OF INVENTION: PATENTS AND COPYRIGHTS IN AMERICAN ECONOMIC DEVELOPMENT, 1790-1920 (NY: Cambridge University Press 2005).

KINGSTON, WILLIAM, INNOVATION, CREATIVITY AND LAW (Dordrecht: Kluwer Academic Publishers 1990).

KNOPPER, STEVE, APPETITE FOR SELF-DESTRUCTION: THE SPECTACULAR CRASH OF THE RECORD INDUSTRY IN THE DIGITAL AGE (NY: Free Press 2009).

Koelman, Kamiel J., *A Hard Nut to Crack: The Protection of Technological Measures*, 22 EIPR 272 (2000).

KOMESAR, NEIL, IMPERFECT ALTERNATIVES: CHOOSING INSTITUTIONS IN LAW, ECONOMICS AND PUBLIC POLICY (Chicago & London: The University of Chicago Press 1994).

Koren, Niva Elkin, *Making Room for Consumers Under the DMCA*, 22 BERKELEY TECH. L.J. 1119 (2007).

KWALL, ROBERTA ROSENTHAL, THE SOUL OF CREATIVITY: FORGING A MORAL RIGHTS LAW FOR THE UNITED STATES (California: Stanford University Press 2010).

Kwall, Roberta Rosenthal, *Copyright and the Moral Right: Is an American Marriage Possible?* 38 VAND. L. REV. 1 (1985).

LANDES, M. WILLIAM & POSNER, A. RICHARD, THE ECONOMIC STRUCTURE OF INTELLECTUAL PROPERTY LAW (Cambridge: The Belknap Press of Harvard University Press 2003).

Landes, William M & Posner, Richard A., *An Economic Analysis of Copyright Law*, 18 J. LEG. STUD. 325 (1989).

LANDSBURG, STEVEN E., PRICE THEORY AND APPLICATIONS (7th ed., Thomson South-Western 2008).

Lange, David, *At Play in the Fields of the Word: Copyright and Construction of Authorship in the Post-Literate Millennium*, 55 L. & CONTEMP. PROBS. 139 (1992).

LATHROP, TAD, THIS BUSINESS OF GLOBAL MUSIC MARKETING (NY: Billboard Books 2007).

LEAFFER, MARSHALL, UNDERSTANDING COPYRIGHT LAW (2nd ed., NY: Matthew Bender 1995).

Lemley, Mark, *Property, Intellectual Property, and Free Riding*, 83 TEXAS L. REV. 1031 (2005).

LESSIG, LAWRENCE, CODE VERSION 2.0 (NY: Basic books 2006).

LESSIG, LAWRENCE, FREE CULTURE: THE NATURE AND FUTURE OF CREATIVITY (NY: Penguin books 2004).

LESSIG, LAWRENCE, THE FUTURE OF IDEAS: THE FATE OF THE COMMONS IN A CONNECTED WORLD (NY: Random House, Inc., 2001).

Leval, Pierre N., *Nimmer Lecture: Fair Use Rescued*, 44 UCLA L. REV. 1449 (1997).

Leval, Pierre N., *Toward a Fair Use Standard*, 103 HARV. L. REV. 1105 (1990).

Leval, Pierre N., *Fair Use or Foul*, 36 J. COPYRIGHT SOC'Y U.S.A. 167 (1989).

LEWINSKI, SILKE, INTERNATIONAL COPYRIGHT LAW AND POLICY (New York: Oxford University Press 2008).

Lewis, Tracy R. & Yao, Dennis A., *Some Reflections on the Antitrust Treatment of Intellectual Property*, 63 ANTITRUST L. J. 603 (1995).

Li Chi, Xingzheng guanli guocheng zhong duotou zhifa de jiejue lujing tanxi [*On the Solutions to Multi-headed Enforcement in the Administrative Process*], 7 Fazhi yu shehui [LEGAL SYSTEM AND SOCIETY] 198 (2009).

Litman, Jessica, *Lawful Personal Use*, 85 TEX. L. REV. 1871(2007).

Litman, Jessica, *Creative Reading*, 70 LAW & CONTEMP. PROB. 175 (2007).

Litman, Jessica, *The Sony Paradox*, 55 CASE W. RES. L. REV. 917(2005).

LITMAN, JESSICA, DIGITAL COPYRIGHT (NY: Prometheus Books 2001).

Litman, Jessica, *The Public Domain*, 39 EMORY L.J. 965 (1990).

Litman, Jessica, *Copyright Legislation and Technological Change*, 68 OREGON LAW REVIEW 275 (1989).

Liu, Joseph P., *Copyright Law's Theory of the Consumer*, 44 B.C. L. REV 397 (2003).

Li Yongming & Lü Yilin, Lun zhishi chanquan zhi gongquan xingzhi – dui "zhishi chanquan shuyu siquan" de buchong [*Public Authority of Intellectual Property*], 4 Zhejiang daxue xuebao [JOURNAL OF ZHEJIANG UNIVERSITY] 60 (2004).

Li Yufeng, Qiangkou xia de falü: Zhongguo banquan shi yanjiu [LAW AT GUNPOINT: RESEARCH ON THE CHINESE COPYRIGHT HISTORY] (Beijing: Zhishi chanquan chubanshe [Intellectual Property Publishing Company] 2006).

Liang Xingguo, Kala OK banquan shoufei zhi zheng yu zhishi chanquan zhong de gongquan wenti [*Public Rights Issues of Intellectual Property in the Disputes of KTV*], 1 Shanghai caijing daxue xuebao [JOURNAL OF SHANGHAI UNIVERSITY OF FINANCE AND ECONOMIC] 46 (2008).

Lin Yifu, Zhidu, jishu yu zhongguo nongye fazhan [SYSTEMS, TECHNOLOGY AND THE AGRICULTURE DEVELOPMENT IN CHINA] (Shanghai: Shanghai San lian shudian, Shanghai renmin chubanshe [Shanghai San Lian Bookstore & Shanghai Renmin Publishing House] 1994).

Liu Chunmao ed., Zhishi chanquan yuanli [FOUNDATIONS OF IP] (Beijing: Zhishi chanquan chubanshe [Intellectual Property Publishing Company] 2002).

Liu Chuntian ed., Zhishi chanquan fa [INTELLECTUAL PROPERTY LAW] (Beijing: Zhongguo renmin daxue chubanshe [China Renmin University Press] 2000).

Liu Songshan, Guojia lifa sanshi nian de huigu yu zhanwang [*Reflection and Prospects of China's Legislation in the Past Thirty Years*], 1 Zhongguo faxue [CHINA LEGAL SCIENCE] 31 (2009).

Liu Wenjing, WTO guize guonei shishi de xingzheng fa wenti [ADMINISTRATIVE LAW ISSUES IN THE IMPLEMENTATION OF WTO

REGULATIONS] (Beijing: Beijing daxue chubanshe [Peking University Press] 2004).

Liu Xingcheng, Zhongguo fan daoban de pingjing [*The Bottleneck of Anti-piracy in China*], 7 Faren [FAREN MAGAZINE] 43 (2009).

LOEWENSTEIN, JOSEPH, THE AUTHOR'S DUE: PRINTING AND THE PREHISTORY OF COPYRIGHT (Chicago: The University of Chicago Press 2002).

Lohmann, Fred, *Measuring the Digital Millennium Copyright Act Against the Darknet: Implication for the Regulation of Technological Protection Measures*, 24 LOYOLA OF LOS ANGELES ENTERTAINMENT LAW REVIEW 635 (2004).

Long Wenmao, Zhishi chanquan fazhexue chulun [SURVEY ON THE LEGAL PHILOSOPHY OF INTELLECTUAL PROPERTY LAW] (Beijing: Renmin chubanshe [People's Publishing House] 2003).

Lu Haijun, Lun heli shiyong zhidu de lifa moshi [*The Legislative Model of the Fair Use System*], 3 Fa shang yanjiu [LAW & COMMERCIAL INDUSTRY RESEARCH] 24 (2007).

Lu Haijun, Heli shiyong yiban tiaokuan de caixiang yu zhengming: heli shiyong zhidu lifa moshi tantao [*General Clause of Fair Use: Research on the Legislative Model of Fair Use*], 2 Zhengfa Luncong [JOURNAL OF POLITICAL SCIENCE AND LAW] 42 (2007).

Lunney, Glynn S., *Reexamining Copyright's Incentives-Access Paradigm*, 49 VAND. L. REV. 483 (1996).

MACMILLAN, FIONA ED., NEW DIRECTIONS IN COPYRIGHT LAW (Cheltenham: Edward Elgar 2005).

MACQUEEN, HECTOR ET AL, CONTEMPORARY INTELLECTUAL PROPERTY: LAW AND POLICY (NY: Oxford University Press 2008).

Madison, Michael J., *Fair Use and Social Practices*, in INTELLECTUAL PROPERTY AND INFORMATION WEALTH: ISSUES AND PRACTICES IN THE DIGITAL AGE, VOL. 1, COPYRIGHT AND RELATED RIGHTS (Peter K. Yu ed., Connecticut: Praeger Publishers 2007).

Madow, Michael P., *Private Ownership of Public Image: Popular Culture and Publicity Rights*, 81 CAL. L. REV. 125 (1993).

MAKEEN, MAKEEN F., COPYRIGHT IN A GLOBAL INFORMATION SOCIETY: THE SCOPE OF COPYRIGHT PROTECTION UNDER INTERNATIONAL, US, UK AND FRENCH LAW (Hague: Kluwer Law International 2000).

Ma Li, "Daoban jia" beihou de xuanji [*Secret behind the Price of Piracy*], 8 Faren [FAREN MAGAZINE] 52 (2009).

MAZZIOTTI, GIUSEPPE, EU DIGITAL COPYRIGHT LAW AND THE END-USER (Heidelberg: Springer 2008).

Merges, Robert P., *The End of Friction? Property Rights and Contract in the "Newtonian" World of On-Line Commerce*, 12 BERKELEY TECH. L.J. 115 (1997).

Merges, Robert P., *Contracting Into Liability Rules: Intellectual Property Rights and Collective Rights Organizations*, 84 CAL. L. REV. 1293 (1996).

Merrill, Thomas, *Capture Theory and the Courts: 1967-1983*, 72 CHI.-KENT L. REV. 1039 (1997).

MERTHA, ANDREW C., THE POLITICS OF PIRACY: INTELLECTUAL PROPERTY IN CONTEMPORARY CHINA (Ithaca: Cornell University Press 2005).

Meuer, Michael, *Too Many Markets or Too Few? Copyright Policy Toward Shared Work*, 77 S. CAL. L. REV. 903 (2004).

Meuer, Michael, *Vertical Restraints and Intellectual Property Law: Beyond Antitrust*, 87 MINN. L. REV. 1871 (2003).

Meurer, Michael J., *Price Discrimination, Personal Use and Piracy: Copyright Protection of Digital Works*, 45 BUFF. L. REV. 845 (1997).

Michelman, Frank I., *Property, Utility, and Fairness: Comments on the Ethical Foundations of "Just Compensation" Law*, 80 HARV. L. REV. 1165 (1967).

MILLER, ARTHUR R. & DAVIS, MICHAEL H., INTELLECTUAL PROPERTY: PATENTS, TRADEMARK, AND COPYRIGHT IN A NUTSHELL (3d ed., US: Thomson/West 2000).

MILLER, MICHAEL, CLOUD COMPUTING: WEB-BASED APPLICATIONS THAT CHANGE THE WAY YOU WORK AND COLLABORATE ONLINE (US: Que Publishing 2009)

Monta, Rudolf, *The Concept of "Copyright" Versus the "Droit d'Auteur"*, 32 S. CAL. L. REV. 177 (1959).

Mulligan, Deirdre K. & Perzanowski, Aaron K., *The Magnificence of the Disaster: Reconstructing the Sony BMG Rootkit Incident*, 22 BERKELEY TECH. L. J. 1157 (2007).

Naftel, Mark & Spiwak, Lawrence J., THE TELECOMS TRADE WAR: THE UNITED STATES, THE EUROPEAN UNION, AND THE WORLD (Oregon: Hart Publishing 2000).

NAGEL, ERNST, THE STRUCTURE OF SCIENCE (London: Routledge & Kegan Paul 1961).

NAGEL, THOMAS, THE POSSIBILITY OF ALTRUISM (Oxford: Clarendon Press 1970).

NATIONAL STUDIES ON ASSESSING THE ECONOMIC CONTRIBUTION OF THE COPYRIGHT-BASED INDUSTRIES (Geneva, Switzerland: World Intellectual Property Organization 2006).

NEEF, DALE, SIESFELD, G. ANTHONY & CEFOLA JACQUELYN EDS., THE ECONOMIC IMPACT OF KNOWLEDGE (Boston: Butterworth Heinemann 1998).

NETANEL, NEIL WEINSTOCK, COPYRIGHT'S PARADOX (NY: Oxford University Press 2008).

Netanel, Neil Weinstock, *Why Has Copyright Expanded? Analysis and Critique,* in NEW DIRECTIONS IN COPYRIGHT LAWS (Fiona Macmillan ed., Vol.6, UK: Edward Elgar 2007).

Netanel, Neil Weinstock, *Impose a Noncommercial Use Levy to Allow Free Peer-to-peer File Sharing*, 17 HARV. J. L. & TECH 1 (2003).

Netanel, Neil Weinstock *Locating Copyright Within the First Amendment Skein*, 54 STAN. L. REV. 1(2001).

Netanel, Neil Weinstock, *Copyright and a Democratic Civil Society*, 106 YALE L.J. 283 (1996).

Netanel, Neil Weinstock, *Alienability Restrictions and the Enhancement of Author Autonomy in United States and Continental Copyright Law*, 12 CARDOZO ARTS & ENT. L. J. 1 (1994).

Nimmer, David, *How Much Solicitude for Fair Use Is There in the Anti-Circumvention Provision of the Digital Millennium Copyright Act?* in THE COMMODIFICATION OF INFORMATION (Niva Elkin-Koren & Neil Weinstock Netanel eds., The Hague: Kluwer Law International 2002).

Nimmer, David, *Does Copyright Abridge the First Amendment Guarantees of Free Speech and Press?* 17 U.C.L.A.L.REV. 1180 (1970).

NIMMER, MELVILLE B. & NIMMER, DAVID, NIMMER ON COPYRIGHT (NY: Matthew Bender 1997).

Nimmer, Melville B., *New Technology and the Law of Copyright: Reprography and Computers*, 15 UCLA L. REV. 931 (1968).

Okediji, Ruth, *Toward an International Fair Use Doctrine*, 39 COLUM. J. TRANSNAT'L L. 75 (2000).

O'Rourke, Maureen A., *Fencing Cyberspace: Drawing Borders in a Virtual World*, 82 MINN. L. REV. 609 (1998).

OSHANA, MARINA, PERSONAL AUTONOMY IN SOCIETY (England: Ashgate Publishing Limited 2006).

PATRY, WILLIAM F. THE FAIR USE PRIVILEGE IN COPYRIGHT LAW (Washington, D.C.: The Bureau of National Affairs, Inc. 1985).

PATTERSON, L. RAY & LINDBERG, STANLEY W., THE NATURE OF COPYRIGHT: A LAW OF USERS' RIGHTS (Athens: The University of Georgia Press 1991).

PATTERSON, LYMAN RAY: COPYRIGHT IN HISTORICAL PERSPECTIVE (Nashville: Vanderbilt University Press 1968).

PETTIT, PHILIP, JUDGING JUSTICE: AN INTRODUCTION TO CONTEMPORARY POLITICAL PHILOSOPHY (London and Boston: Routledge & Kegan Paul 1980).

Peukert, Alexander, *A Bipolar Copyright System for the Digital Network Environment*, in PEER-TO-PEER FILE SHARING AND SECONDARY LIABILITY IN COPYRIGHT LAW (Alain Strowel ed., UK: Edward Elgar Publishing Limited 2009).

PFLEEGER, CHARLES P. & PFLEEGER, SHARI LAWRENCE, SECURITY IN COMPUTING (3rd ed., New Jersey: Pearson Education, Inc 2003).

PHILIPS, LOUIS, THE ECONOMICS OF PRICE DISCRIMINATION (NY: Cambridge University Press 1981).

PORSDAM, HELLE ED., COPYRIGHT AND OTHER FAIRY TALES: HANS CHRISTIAN ANDERSEN AND THE COMMODIFICATION OF CREATIVITY (UK: Edward Elgar 2006).

Post, Robert, *Reconciling Theory and Doctrine in First Amendment Jurisprudence*, 88 CAL. L. REV. 2353 (2000).

PUGATCH, MEIR PEREZ ED., THE INTELLECTUAL PROPERTY DEBATE: PERSPECTIVES FROM LAW, ECONOMICS AND POLITICAL ECONOMY (UK: Edward Elgar 2006).

PUTNAM, GEORGE H., BOOKS AND THEIR MAKERS IN THE MIDDLE AGES (NY: Hillary House Publishers 1962).

Qiao Sheng, Xinxi wangluo chuanbo quan lifa pingjia yu wanshan [*Evaluation of the Lawmaking Process of the Information Network Transmission Right and Its Improvement*], 4 Zhongguo faxue [CHINESE LEGAL SCIENCE] 15 (2004).

Quanguo Renda Changweihui Fazhi Gongzuo Weiyuanhui Minfashi [Cilvil Law Office, Law Committee of the Standing Committee of the National People's Congress] ed., Zhonghua Renmin Gongheguo zhuzuoquan fa xiugai lifa ziliao xuan [SELECTED LEGISLATIVE MATERIALS OF THE CHINESE COPYRIGHT LAW] (Beijing: Falü chubanshe [Law Press China] 2002).

RAHNASTO, ILKKA, INTELLECTUAL PROPERTY RIGHTS, EXTERNAL EFFECTS AND ANTI-TRUST LAW: LEVERAGING IPRS IN THE COMMUNICATIONS INDUSTRY (NY: Oxford University Press 2003).

RATHBONE, ANDY, PCS: THE MISSING MANUAL, (CA: O'Reilly Media, Inc. 2006).

Reese, R. Anthony, *The First Sale Doctrine in the Era of Digital Networks*, 44 B. C. L. REV. 577 (2003).

Reichman, Jerome H. & Franklin, Jonathan A., *Privately Legislated Intellectual Property Rights: Reconciling Freedom of Contract with Public Good Uses of Information*, 147 U. PENN. L. REV. 875 (1999).

REINBOTHE, JORG & LEWINSKI, SILKE VON, THE WIPO TREATIES 1996 (London: Butterworths Lexis Nexis 2002).

RHEINGOLD, HOWARD, SMART MOBS: THE NEXT SOCIAL REVOLUTION (Cambridge, MA: Perseus Publishing 2002).

RICKETSON, SAM & GINSBURG, JANE, INTERNATIONAL COPYRIGHT AND NEIGHBORING RIGHTS: THE BERNE CONVENTION AND BEYOND (NY: Oxford University Press 2006).

RIMMER, MATTHEW, DIGITAL COPYRIGHT AND THE CONSUMER REVOLUTION: HANDS OFF MY IPOD (UK: Edward Elgar 2007).

RITTINGHOUSE, JOHN W. & RANSOME, JAMES F., CLOUD COMPUTING: IMPLEMENTATION, MANAGEMENT, AND SECURITY (Boca Raton: CRC Press 2010).

R. MacFarquhar & Fei Zhengqing eds., Jianqiao Zhonghua Renmin Gongheguo shi [THE CAMBRIDGE HISTORY OF CHINA] (Beijing: Zhongguo shehui kexue chubanshe [China Social Sciences Publishing House] 1998).

Roffe, Pedro, Spennemann, Christoph & Braun, Johanna von, *Intellectual Property Rights in Free Trade Agreements: Moving Beyond TRIPS Minimum Standards*, in RESEARCH HANDBOOK ON THE PROTECTION OF IP UNDER WTO RULES (Carlos M. Correa ed., UK: Edward Elgar Publishing Limited 2010).

ROONEY, DAVID, HEARN, GREG, MANDEVILLE, THOMAS & JOSEPH, RICHARD, PUBLIC POLICY IN KNOWLEDGE-BASED ECONOMIES: FOUNDATIONS AND FRAMEWORKS (UK: Edward Elgar Publishing Limited 2003).

ROSE, MARK, AUTHORS AND OWNERS: THE INVENTION OF COPYRIGHT (Cambridge: Harvard University Press 1993).

Rubenfeld, Jed, *The Freedom of Imagination: Copyright's Constitutionality*, 112 YALE L. J. 1 (2002).

RUSHING, FRANCIS W., & BROWUYN, CAROLE G. EDS., INTELLECTUAL PROPERTY RIGHTS IN SCIENCE, TECHNOLOGY, AND ECONOMIC PERFORMANCE: INTERNATIONAL COMPARISONS (Boulder: Westview Press 1990).

SAINT-AMOUR, PAUL K., THE COPYWRIGHTS: INTELLECTUAL PROPERTY AND THE LITERARY IMAGINATION (NY: Cornell University Press 2003).

Sagi, Guy, *The Oligopolistic Pricing Problem: A Suggested Price Freeze Remedy*, 2008 COLUM. BUS. L. REV. 269 (2008).

SAMET, ROBERT H., LONG-RANGE FUTURES RESEARCH: AN APPLICATION OF COMPLEXITY SCIENCE (US: BookSurge Publishing 2008).

Samuelson, Pamela et al., *The Copyright Principles Project: Directions for Reform*, 25 BERKELEY TECH. L. J. 1 (2010).

SANDEL, MICHAEL J., JUSTICE: WHAT'S THE RIGHT THING TO DO? (London: Penguin Books 2009).

Schneck, Paul B., *Persistent Access Control to Prevent Privacy of Digital Information*, 87 PROCEEDINGS OF THE IEEE 1239 (1999).

SELL, SUSAN K., PRIVATE POWER, PUBLIC LAW: THE GLOBALIZATION OF INTELLECTUAL PROPERTY RIGHTS (Cambridge: Cambridge University Press 2003).

SENFTLEBEN, MARTIN, COPYRIGHT, LIMITATIONS AND THE THREE-STEP TEST: AN ANALYSIS OF THE THREE-STEP TEST IN INTERNATIONAL AND EC COPYRIGHT LAW (Hague: Kluwer Law International 2004).

SEVILLE, CATHERINE, THE INTERNATIONALISATION OF COPYRIGHT LAW: BOOKS, BUCCANEERS AND THE BLACK FLAG IN THE NINETEENTH CENTURY (Cambridge: Cambridge University Press 2006).

SEVILLE, CATHERINE, LITERARY COPYRIGHT REFORM IN EARLY VICTORIAN ENGLAND: THE FRAMING OF THE 1842 COPYRIGHT ACT (Cambridge: Cambridge University Press 1999).

Shen Rengan et. al. eds., Zhonghua Renmin Gongheguo zhuzuoquan fa jianghua [AN INTRODUCTION TO THE CHINESE COPYRIGHT LAW] (Beijing: Falü chubanshe [Law Press China] 1991).

Shen Rengan, Banquan fa yu banquan baohu de shehui zuoyong [*Social Function of Copyright Law and Copyrihgt Protection*], in Banquan yanjiu wenxuan [SELECTED WORKS OF COPYRIGHT RESEARCH] (Zhongguo banquan yanjiuhui [Copyright Society of China] ed., Beijing: Shangwu yinshuguan [The Commercial Press] 1995).

Shen Rengan, Guanyu dui zhuzuoquan de xianzhi [*On the Restrictions of Copyright*], in Jiang Ping et al., Zhonghua Renmin Gongheguo zhuzuoquan fa jiangxi [ANALYSIS ON THE CHINESE COPYRIGHT LAW] (Beijing: Zhongguo guoji guangbo chubanshe [China International Broadcasting Press] 1991).

Shen Rengan & Zhong Yingke, Banquan fa qianshuo [ELEMENTARY INTRODUCTION OF THE COPYRIGHT LAW] (Beijing: Falü chubanshe [Law Press China] 1982).

Shen, Yuanyuan, *Conceptions and Receptions of Legality: Understanding the Complexity of Law Reform in Modern China*, in THE LIMITS OF THE RULE OF LAW IN CHINA (Karen G. Turner, James V. Feinerman, & R. Kent Guy eds., University of Washington Press 2000).

SHERMAN, BRAD & BENTLY, LIONEL, THE MAKING OF MODERN INTELLECTUAL PROPERTY LAW: THE BRITISH EXPERIENCE 1760-1911 (Cambridge: Cambridge University Press 1999).

SHERMAN, BRAD & STROWEL, ALAIN EDS., OF AUTHORS AND ORIGINS: ESSAYS ON COPYRIGHT LAW (Oxford: Clarendon Press 1994).

Solberg, Thorvald compiles, COPYRIGHT ENACTMENTS OF THE UNITED STATES: 1783-1906 (Washington, D.C.: Library of Congress 1906).

Song Muwen, Guanyu woguo zhuzuoquan fa de xiugai [*Concerning the Revision of the Chinese Copyright Law*], in 1 Chuban kexue [PUBLISHING SCIENCE] 4 (2002).

SOTO, JESÚS HUERTA DE, THE THEORY OF DYNAMIC EFFICIENCY (Oxon: Routledge 2009).

Stadler, Sara K., *Copyright as Trade Regulation*, 155 U. PA. L. REV. 899 (2007).

Stadler, Sara K., *Incentives and Expectation in Copyright*, 58 HASTINGS L. J. 433, (2007).

STAMATOUDI, IRINI A., COPYRIGHT AND MULTIMEDIA PRODUCTS: A COMPARATIVE ANALYSIS (Cambridge University Press 2002).

Sterk, Stewart E., *Rhetoric and Reality in Copyright Law*, 94 MICH. L. REV. 1197 (1996).

STERLING, ADRIAN, WORLD COPYRIGHT LAW (London: Sweet & Maxwell 1999).

Sterling, J.A.L., *Creator's Right and the Bridge between Author's Right and Copyright*, 29 IIC 302 (1998).

Strowel, Alain, *Droit d'Auteur and Copyright: Between History and Nature*, in OF AUTHORS AND ORIGINS: ESSAYS ON COPYRIGHT LAW (Brad Sherman & Alain Strowel eds., Oxford: Clarendon Press 1994).

Su Li, Songfa xiaxiang [SENDING THE LAW TO THE COUNTRYSIDE] (Beijing: Zhongguo zhengfa daxue chubanshe [China University of Political Science and Law Press] 2000).

Su Li, Zhidu shi ruhe xingcheng de [HOW A SYSTEM IS FORMULATED] (Guangzhou: Zhongshan daxue chubanshe [Sun Yat-sen University Press] 1999)

Su Li, Fazhi jiqi bentu ziyuan [THE NATIVE RESOURCE OF RULE OF LAW] (Beijing: Zhongguo zhengfa daxue chubanshe [China University of Political Science and Law Press] 1996).

Sunder, Madhavi, *IP3*, 59 STANFORD L. REV. 257 (2006).

Tan Ying, Ruanjian daoban chengyin yu zhili jizhi fenxi [*Analysis of the Reasons for Software Piracy and Governance Mechanisms*], 1 Ruanjian daokan [SOFTWARE GUIDE] 3 (2009).

TAYLOR, JIM, DVD DEMYSTIFIED (2d ed. NY: McGraw-Hill Companies, Inc. 2000).

Tehranian, John, *Et Tu, Fair Use? The Triumph of Natural-Law Copyright*, 38 U.C. DAVIS L. REV. 465(2005).

TIAN, YIJUN, RE-THINKING INTELLECTUAL PROPERTY: THE POLITICAL ECONOMY OF COPYRIGHT IN THE DIGITAL ERA (NY: Routledge-Cavendish 2009).

Tokaji, Daniel P., *First Amendment Equal Protection: On Discretion, Inequality, and Participation*, 101 MICH. L. REV. 2409 (2003).

TORREMANS, PAUL L.C. ED., COPYRIGHT AND HUMAN RIGHTS: FREEDOM OF EXPRESSION – INTELLECTUAL PROPERTY – PRIVACY (The Hague: Kluwer Law International 2004).

TOWSE, RUTH ED., COPYRIGHT IN THE CULTURAL INDUSTRIES (UK: Edward Elgar 2002).

Turner, Karen G., Feinerman, James V., & Guy, R. Kent eds., THE LIMITS OF THE RULE OF LAW IN CHINA (University of Washington Press 2000).

Tushnet, Rebecca, *Copy This Essay: How Fair Use Doctrine Harms Free Speech and How Copying Serves It*, 114 YALE L. J. 535 (2004).

Tushnet, Rebecca, *Copyright as a Model for Free Speech Law: What Copyright Has in Common with Anti-Pornography Laws, Campaign Finance Reform, and Telecommunications Regulation*, 42 B. C. L. REV. 1 (2000).

Tussey, Deborah, *From Fan Sites to Filesharing: Personal Use in Cyberspace*, 35 GA. L. REV. 1129 (2001).

VAIDHYANATHAN, SIVA, THE ANARCHIST IN THE LIBRARY: HOW THE CLASH BETWEEN FREEDOM AND CONTROL IS HACKING THE REAL WORLD AND CRASHING THE SYSTEM (NY: Basic Books 2004).

Valkonen, Sami J. & White, Lawrence J., *An Economic Model for the Incentive/Access Paradigm of Copyright Propertization: An Argument in Support of the Orphan Works Act*, 29 HASTINGS COMM. & ENT. L. J. 359(2007).

VAVER, DAVID, COPYRIGHT LAW (Toronto: Irwin Law 2000).

VELTE, ANTHONY T., VELTE, TOBY J. & ELSENPETER, ROBERT, CLOUD COMPUTING: A PRACTICAL APPROACH (US: McGraw-Hill Companies 2010).

VERMEULE, ADRIAN, JUDGING UNDER UNCERTAINTY: AN INSTITUTIONAL THEORY OF LEGAL INTERPRETATION (Cambridge: Harvard University Press 2006).

VERPEAUX, MICHEL, FREEDOM OF EXPRESSION: IN CONSTITUTIONAL AND INTERNATIONAL CASE LAW (Strasbourg: Council of Europe Publishing 2010).

Vinje, Thomas C., *Copyright Imperilled?* 21 EUR. INTELL. PROP. REV. 192 (1999).

Volokh, Eugene, *Freedom of Speech and Intellectual Property: Some Thoughts After Eldred, 44 Liquormart, and Bartnicki*, 40 HOUS. L. REV. 697 (2003).

Wagner, R. Polk, *Information Wants to Be Free: Intellectual Property and the Mythologies of Control*, 103 COLUM. L. REV. 995 (2003).

Wang Chunyan, Lun zhuzuoquan qinquan xingwei de tedian jiqi rending [*Characteristics and Determination of Copyright Infringement*], 2 Zhongwai faxue [PEKING UNIVERSITY LAW JOURNAL] 39 (1996).

Wang Lanping, Jindai zhongguo zhuzuoquan fa de chengzhang (1903-1910) [DEVELOPMENT OF THE COPYRIGHT LAW IN NEOTERIC CHINA: 1903-1910] (Beijing: Beijing daxue chubanshe [Peking University Press] 2006).

Wang Lanping, Zhongguo fazhi jindaihua guocheng zhong de sanbu zhuzuoquan fa [*Three Copyright Laws in the Process of Modernization of Chinese Law*], 3 Bijiao fa yanjiu [JOURNAL OF COMPARATIVE LAW] 44 (2005).

Wang Qian, Lun wangluo huanjing zhong de "shouci xiaoshou yuanze" [*"Principles of First Sale" in the Internet Environment*], 3 Faxue zazhi [LAW SCIENCE MAGAZINE] 117 (2006).

Wang Qian, P2P ruanjian zuizhong yonghu banquan qinquan wenti yanjiu [*Research on End User Liability with P2P Software*], 5 Zhishi chanquan [INTELLECTUAL PROPERTY] 9 (2004).

Wang Qian, Dui jishu cuoshi lifa baohu de bijiao yanjiu [*Comparative Study on the Protection of Technological Protection Measures*], 2 Zhishi chanquan [INTELLECTUAL PROPERTY] 3 (2003).

Wang Taiping & Li Changhao, Zhuzuoquan baohu de shuangchong xianzhi –yi sihai gujuan an yin zhi wei jian [*Dual Restriction of Copyright Protection: A Critical Evaluation of the Dead Sea Ancient Literature Case*], 4 Zhishi chanquan [INTELLECTUAL PROPERTY] 58 (2007).

Wang Yingming, Shichang jingji ying zhuzhong dui zhishi chanquan de xingzheng baohu [*The Market Economy Needs More Emphasis on the Administrative Protection of IPRs*], 11 Xingzheng yu fa [PUBLIC ADMINISTRATION & LAW] 78 (2008).

WATT, RICHARD, COPYRIGHT AND ECONOMIC THEORY: FRIENDS OR FOES? (UK: Edward Elgar 2000).

Weinberg, Jonathan, *Hardware-Based ID, Rights Management, and Trusted Systems*, 52 STAN. L. REV. 1251 (2000).

Wei Tao, Daoban xianxiang de jingji genyuan tanxi [*Research on the Economic Root of Piracy*], 14 Jinrong jingji [FINANCE & ECONOMY] 120 (2009).

Wei Zhi, Cong bo'er'ni gongyue kan zhongguo zhuzuoquan fa zhi xiuding [*An Evaluation on the Revision of the Chinese Copyright Law from the Berne Convention Perspective*], 5 Zhongwai faxue [PEKING UNIVERSITY LAW JOURNAL] 91(1997).

WENDLER, MICHAEL, TREMML, BERND, & BUECKER, BERNARD EDS., KEY ASPECTS OF GERMAN BUSINESS LAW: A PRACTICAL MANUAL (4th ed., Verlag: Springer 2008).

WONG, KENNY K. S. & LEE, ALICE S. C., A PRACTICAL APPROACH TO INTELLECTUAL PROPERTY LAW IN HONG KONG (Hong Kong: Sweet & Maxwell Asia 2002).

Wu Handong, Guanyu Zhongguo zhuzuoquan fa guannian de lishi sikao [*A Historical Review of the Chinese Copyright Law Notion*], 3 Fa shang yanjiu [LAW & COMMERCIAL INDUSTRY RESEARCH] 44 (1995).

Wu Handong, Zhuzuoquan heli shiyong zhidu yanjiu [RESEARCH ON THE FAIR USE OF COPYRIGHT] (Beijing: Zhongguo zhengfa daxue chubanshe [China University of Political Science and Law Press] 1996).

Wu Shuchen, falü wenhua yu Zhongguo chuantong falü wenhua lungang [*Outline of Legal Culture and the Chinese Traditional Legal Culture*], in Ruxue yu falü wenhua [CONFUCIANISM AND LEGAL CULTURE] (Zhongguo ruxue yu falü wenhua yanjiuhui [China Confucianism and Legal Culture Seminar] ed., Shanghai: Fudan daxue chubanshe [Fudan University Press] 1992).

Wyld, David C., *Risk in the Clouds?: Security Issues Facing Government Use of Cloud Computing*, in INNOVATIONS IN COMPUTING SCIENCES AND SOFTWARE ENGINEERING (Tarek Sobh & Khaled Elleithy eds., Heidelberg: Springer 2010).

Xiao Xun, Cong woguo de shiji chufa jiejue banquan (zhuzuoquan) lifa zhong you zhengyi de wenti [*Working Solution for Disputed Issues in Copyright Lawmaking*], in Banquan yanjiu wenxuan [SELECTED WORKS OF COPYRIGHT RESEARCH] (Zhongguo banquan yanjiuhui [Copyright Society of China] ed., Beijing: Shangwu yinshuguan [The Commercial Press] 1995).

Yang Chongsen, Zhuzuoquan zhi baohu [PROTECTION OF COPYRIGHT] (Taiwan: Taiwan zhengzhong shuju [Taiwan Zhengzhong Publishing House] 1977).

Yang Hongjun, Banquan qiangzhi xuke zhidu lun [*The Compulsory License System of Copyright*], 4 Zhishi chanquan [INTELLECTUAL PROPERTY] 30 (2008).

Ying Songnian, Yifa xingzheng lungang [*On the Administration by Law*], 1 Zhongguo faxue [CHINESE LEGAL SCIENCE] 29 (1997).

Yonay, Guy, *Book Note: To Steal A Book Is an Elegant Offense: Intellectual Property Law in Chinese Civilization,* 8 HARV. J. L. & TECH. 537 (1995).

Yoo, Christopher S., *Copyright and Product Differentiation*, 79 N. Y. U. L. REV. 212 (2004).

Yoshiyuki Tamura, Zhishi chanquan fa de lilun [*The Theory of Intellectual Property Law*], in Zhishi chanquan niankan [INTELLECTUAL PROPERTY RIGHTS ANNUAL JOURNAL] (initial issue, Li Daodao trans., Beijing: Beijing daxue chubanshe [Peking University Press] 2005).

Yu, Peter K., *From Pirates to Partners (Episode II): Protecting Intellectual Property in Post- WTO China*, 55 AM. U. L. REV, 901(2006).

Yu, Peter K., *Four Common Misconceptions about Copyright Piracy*, 26 LOY. L.A. INT'L & COMP. L. REV. 127 (2003).

Yu Yu & Ji Xiaoxin, Woguo zhuzuoquan heli shiyong panduan biaozhun de fansi yu chonggou [*The Reflection and Reconstruction of the Assessment Criterion for the Fair Use of Copyright in China*], 3 Faxue luntan [LEGAL FORUM] 90 (2007).

ZEMER, LIOR, THE IDEA OF AUTHORSHIP IN COPYRIGHT (England: Ashgate Publishing Limited 2007).

Zhang Jin, Banquan fa zhong siren fuzhi wenti yanjiu – cong yinshuaji dao hulianwang [RESEARCH ON PRIVATE COPYING IN COPYRIGHT LAW – FROM THE PRINTING PRESS TO THE INTERNET] (Beijing: Zhongguo zhengfa daxue chubanshe [China University of Political Science and Law Press] 2009).

Zhang Yumin & Li Yufeng, Zhongguo banquan shigang [*A Historical Outline of the Chinese Copyright System*], 1 Keji yu falü [SCIENCE-TECHNOLOGY AND LAW] 42 (2004).

Zhang Yumin ed., Zhishi chanquan faxue [INTELLECTUAL PROPERTY JURISPRUDENCE] (Beijing: Zhongguo jiancha chubanshe [China Procuratorial Press] 2002).

Zhang Yurui, Hulianwang shang zhishi chanquan – susong yu falü [INTELLECTUAL PROPERTY ON THE INTERNET: LITIGATION AND LAW] (Beijing: Renmin fayuan chubanshe [Court Press] 2000).

Zheng Chengsi, Woguo zhuzuoquan fa xiuzheng qianyi [*Comment on the Revision of the Chinese Copyright Law*], 6 Zhuzuoquan [COPYRIGHT] 35 (2001).

Zheng Chengsi, Xinxi chanquan yu banquan fa [*Information Property Right and Copyright Law*], in Banquan yanjiu wenxuan [SELECTED WORKS OF COPYRIGHT RESEARCH] (Zhongguo banquan yanjiuhui [Copyright Society of China] ed., Beijing: Shangwu yinshuguan [The Commercial Press] 1995).

Zheng Chengsi, Zhishi chanquan baohu shiwu quanshu [A PRACTICAL MANUAL OF INTELLECTUAL PROPERTY PROTECTION] (Beijing: Zhongguo yanshi chubanshe [China Yanshi Press] 1995).

Zheng Chengsi ed., Zhishi chanquan anli pingxi [IP CASE REVIEW] (Beijing: Falü chubanshe [Law Press China] 1994).

Zheng Chengsi, Zhishi chanquan fa tonglun [GENERAL SURVEY ON INTELLECTUAL PROPERTY LAW] (Beijing: Falü chubanshe [Law Press China] 1986).

Zhou Lin, Zhongguo banquan shi yanjiu de jitiao xiansuo [*Some Clues for Researching Chinese Copyright History*], in Zhongguo banquan shi yanjiu wenxian [LITERATURE OF CHINESE COPYRIGHT HISTORY] (Zhou Lin & Li Mingshan eds., Beijing: Zhongguo fangzheng chubanshe [China Fangzheng Press] 1999).

Zhou Lin & Li Mingshan eds., Zhongguo banquan shi yanjiu wenxian [LITERATURE OF THE CHINESE COPYRIGHT HISTORY] (Beijing: Zhongguo fangzheng chubanshe [China Fangzheng Press] 1999).

Zhou Yong, Shuzi banquan de shidu baohu jiqi shixian [*Appropriate Protection of Digital Copyright and Its Implementation*], in Wenhua chanye yanjiu zhanlüe yu duice [RESEARCH ON CULTURAL INDUSTRY: STRATEGY AND MEASURE] (Xiong Chengyu et al. eds., Beijing: Qinghua daxue chubanshe [Tsinghua University Press] 2006).

Zhu Xuezhong & Huang Jing, Shilun woguo zhishi chanquan xingzheng guanli jigou de yitihua shezhi [*On the Unified IP Administrative Agency in China*], 3 Keji yu falü [SCIENCE TECHNOLOGY AND LAW] 82 (2004).

Ziebro, Monique & Northcraft, Gregory, *Connecting the Dots: Network Development, Information Flow, and Creativity in Groups*, in CREATIVITY IN GROUPS (Elizabeth A. Mannix, Margaret A. Neale & Jack A. Goncalo eds., UK: Emerald Group Publishing Limited 2009).

ZITTRAIN, JONATHAN, THE FUTURE OF THE INTERNET – AND HOW TO STOP IT (New Haven & London: Yale University Press 2008).

Zittrain, Jonathan, *The Generative Internet*, 119 HARV. L. R. 1974 (2006).

ZUCKERT, RACHEL, KANT ON BEAUTY AND BIOLOGY: AN INTERPRETATION OF THE *CRITIQUE OF JUDGMENT* (UK: Cambridge University Press 2007).

Zuigao Renmin Fayuan Minsan Ting [The Third Civil tribunal of Supreme Court], Zhishi chanquan shenpan zhidao yu cankao [IP TRIAL GUIDANCE AND REFERENCE] (vol.7, Beijing: Falü chubanshe [Law Press China] 2004).

www.ingramcontent.com/pod-product-compliance
Ingram Content Group UK Ltd.
Pitfield, Milton Keynes, MK11 3LW, UK
UKHW021848100926
8556IPUK00020B/1119

9 783631 641880